Lecture Notes in Mathematics

C.I.M.E. Foundation Subseries

Volume 2260

More information about this subseries at http://www.springer.com/series/3114

Fondazione C.I.M.E., Firenze

C.I.M.E. stands for *Centro Internazionale Matematico Estivo*, that is, International Mathematical Summer Centre. Conceived in the early fifties, it was born in 1954 in Florence, Italy, and welcomed by the world mathematical community: it continues successfully, year for year, to this day.

Many mathematicians from all over the world have been involved in a way or another in C.I.M.E.'s activities over the years. The main purpose and mode of functioning of the Centre may be summarised as follows: every year, during the summer, sessions on different themes from pure and applied mathematics are offered by application to mathematicians from all countries. A Session is generally based on three or four main courses given by specialists of international renown, plus a certain number of seminars, and is held in an attractive rural location in Italy.

The aim of a C.I.M.E. session is to bring to the attention of younger researchers the origins, development, and perspectives of some very active branch of mathematical research. The topics of the courses are generally of international resonance. The full immersion atmosphere of the courses and the daily exchange among participants are thus an initiation to international collaboration in mathematical research.

C.I.M.E. Director (2002 – 2014)
Pietro Zecca
Dipartimento di Energetica "S. Stecco"
Università di Firenze
Via S. Marta, 3
50139 Florence
Italy
e-mail: zecca@unifi.it

C.I.M.E. Director (2015 –)
Elvira Mascolo
Dipartimento di Matematica "U. Dini"
Università di Firenze
viale G.B. Morgagni 67/A
50134 Florence
Italy
e-mail: mascolo@math.unifi.it

C.I.M.E. Secretary
Paolo Salani
Dipartimento di Matematica "U. Dini"
Università di Firenze
viale G.B. Morgagni 67/A
50134 Florence
Italy
e-mail: salani@math.unifi.it

CIME activity is carried out with the collaboration and financial support of INdAM (Istituto Nazionale di Alta Matematica)

For more information see CIME's homepage: **http://www.cime.unifi.it**

Antonio DeSimone • Benoît Perthame •
Alfio Quarteroni • Lev Truskinovsky

The Mathematics of Mechanobiology

Cetraro, Italy 2018

Davide Ambrosi • Pasquale Ciarletta

Editors

FONDAZIONE
CIME
ROBERTO CONTI
CENTRO INTERNAZIONALE MATEMATICO ESTIVO
INTERNATIONAL MATHEMATICAL SUMMER CENTER

Authors

Antonio DeSimone
The BioRobotics Institute
Sant'Anna School of Advanced Studies
Pisa, Italy

SISSA-MathLab
International School for Advanced Studies
Trieste, Italy

Alfio Quarteroni
MOX
Mathematics Department
Politecnico di Milano
Italy

Institute of Mathematics
École Polytechnique Fédérale de Lausanne
Lausanne, Switzerland

Benoît Perthame
CNRS, Sorbonne Université
Inria, Université de Paris
Laboratoire Jacques-Louis Lions
Paris, France

Lev Truskinovsky
ESPCI
PMMH
CNRS – UMR 7636 PSL-ESPCI
Paris, France

Editors

Davide Ambrosi
Department of Mathematical Sciences
Polytechnic University of Turin
Torino, Italy

Pasquale Ciarletta (iD)
MOX, Dipartimento Di Matematica
Polytechnic University of Milan
Milano, Italy

ISSN 0075-8434   ISSN 1617-9692   (electronic)
Lecture Notes in Mathematics
C.I.M.E. Foundation Subseries
ISBN 978-3-030-45196-7   ISBN 978-3-030-45197-4   (eBook)
https://doi.org/10.1007/978-3-030-45197-4

Mathematics Subject Classification: 35Q92, 65M60, 70Q05, 92B05, 92C10

This Springer imprint is published by the registered company Springer Nature Switzerland AG.
The registered company address is: Gewerbestrasse 11, 6330 Cham, Switzerland

Preface

In the last couple of decades, the mathematical research at the interface with physics and biology is generating a number of exciting results that are attracting an increasing number of scientists. The emerging need of a theoretical framing of the experimental observations and the search for a suitable formal setting pose a great variety of novel mathematical problems that range from modeling to qualitative analysis, up to numerical methods and rigorous analysis. In particular, Mechanobiology is an emerging research field that stems at the triple point between physics, biology, and mathematics. The large amount of quantitative data that are nowadays made available by the dramatic improvements in imaging technique offer the opportunity to apply mathematical models and methods to understand, to predict, and even to control a number of mechanochemical processes that occur in the biological matter at different scales, from cells to tissues. The mechanics of living matters exhibits a number of peculiar behaviors that challenge classical approaches, driving a number of generalizations of standard frameworks and the revisitation of classical paradigms.

During the last few years, the mathematical community has devoted an increasing interest to the challenges offered by the application of classical methods of continuum mechanics to living systems. The CIME Course "The Mathematics of Mechanobiology" held in Cetraro on August 27–31, 2018 was the occasion to take a snapshot of the state of the art of such a vivid activity. The speakers offered a critical review of the classical and recent results in four specific major areas:

- Cell motility and locomotion by shape control (Antonio DeSimone)
- Models of cell motion and tissue growth (Benoit Perthame)
- Numerical simulation of cardiac electromechanics (Alfio Quarteroni)
- Power-stroke-driven muscle contraction (Lev Truskinovsky)

The lecture notes written by the invited speakers are now collected in this volume of the CIME Lectures collection. We strongly acknowledge the authors for their effort in providing an influential view of key subjects at the core of the research in mathematics for mechanobiology: we believe that this book will be a milestone from now on for all those who are interested in entering this fascinating research field.

The editors are indebted to Ellen Kuhl for her valuable support in the scientific direction of the summer school. Special thanks are due to Elvira Mascolo, Paolo Salani, and Alfredo Marzocchi, in their role of Director, Scientific Secretary, and Member of the CIME Foundation they orchestrated a perfect financial and logistic assistance. We finally wish to acknowledge the financial support of Politecnico di Milano and Politecnico di Torino and the excellent work of all the staff of Grand Hotel San Michele for providing a flawless and very enjoyable stay to all participants.

Torino, Italy   Davide Ambrosi
Milano, Italy   Pasquale Ciarletta

Abstract

This volume collects notes of the lectures delivered at the CIME-EMS Summer School in Applied Mathematics *The Mathematics of Mechanobiology* held in Cetraro, Italy, on August 27–31, 2018.

Contents

Chapter 1
Cell Motility and Locomotion by Shape Control

Antonio DeSimone

Abstract Mathematical modeling and quantitative study of biological motility is producing new biophysical insight and opportunities for discoveries at the level of both fundamental science and technology. One example is the elucidation of how complex behavior of simple organisms emerges from specific (and sophisticated) body architectures, and how this is affected by environmental cues. Moreover, the two-directional interaction between biology and mechanics is promoting new approaches to problems in engineering and in the life sciences: understand biology by constructing bio-inspired machines, build new machines thanks to bio-inspiration.

This article contains an introduction to the mathematical study of swimming locomotion of unicellular organisms (e.g., unicellular algae). We use the tools of geometric control theory to identify some general principles governing life at low Reynolds numbers, that can guide the design of engineered devices trying to replicate the successes of their biological counterparts. Locomotion strategies employed by biological organism are, in fact, a rich source of inspiration for studying mechanisms for shape control. We focus on morphing mechanisms based on Gauss' *theorema egregium*, which shows that the curvature of a thin shell can be controlled through lateral modulations of stretches induced in its mid-surface. We discuss some examples of this Gaussian morphing principle both in nature and technology.

A. DeSimone (✉)
The BioRobotics Institute, Scuola Superiore Sant'Anna, Pisa, Italy

Department of Excellence in Robotics and A.I., Scuola Superiore Sant'Anna, Pisa, Italy

SISSA-MathLab, International School for Advanced Studies, Trieste, Italy
e-mail: a.desimone@santannapisa.it; desimone@sissa.it

© The Editor(s) (if applicable) and The Author(s), under exclusive licence
to Springer Nature Switzerland AG 2020
A. DeSimone et al., *The Mathematics of Mechanobiology*,
Lecture Notes in Mathematics 2260, https://doi.org/10.1007/978-3-030-45197-4_1

1

1.1   Introduction

Motility refers to the ability to move spontaneously. In biology, this is related to the execution of a biological function involving the (active, purposeful) motion of the whole body of an organism or of some of its parts. At the level of individual cells or tissues, motility is crucial in many important biological processes such as cell migration, the immune system response, the establishment of neuron synapses, wound healing, just to name a few. More broadly, motility is fundamental to both the origin of life and the propagation of lethal diseases. Example are the unicellular swimming of sperm cells, the motion of bacteria and parasites in humans, animals and plants, the invasion of nearby tissues by metastatic tumor cells, and the list could continue.

Swimming locomotion of unicellular organisms has provided particularly fertile grounds for the application of mathematical/physical modeling and of quantitative methods to biology. The swimming behavior of micro-organisms has been analyzed through the lenses of mathematical models based on the physical laws of fluid dynamics, and it has attracted considerable attention in the recent biophysical literature.

Among the successes on the biology side, one can list the early discovery of the basic propulsion mechanisms in bacteria (through one rotary motor located at the proximal end of each bacterial flagellum [20]) and flagellated eukaryotes (through molecular motors distributed along the whole length of an eukaryotic flagellum [76]). These have both been discovered through arguments based on physical modeling, when direct observation of the active motors driving flagellar motion was not yet possible. More generally, numerous more recent contributions have established how complex behavior of simple organisms emerges from specific body architectures, and how this is affected by environmental cues. Unicellular organisms represent particularly valuable model systems for the study of the simplest mechanisms of sensing, decision making, and response in biology because they do not involve the intervention of a nervous system (and a brain) in the cascade of regulatory processes.

In addition to the biological motivation to study them, the proficiency exhibited by unicellular swimmers in navigating complex environments, including the human body, has fueled the hope that new bio-inspired biomedical devices can be engineered by trying to learn and replicate the mechanism that work for the biological templates. This includes miniaturized robotic systems for diagnostics, therapeutics, targeted drug delivery, minimally invasive surgery, micro-manipulation, microfluidics. Moreover, the two-directional interaction between biology and mechanics is promoting new approaches to engineering problems and in the life sciences: understand biology by constructing bio-inspired machines, build new machines thanks to bio-inspiration. In these notes, we will try to convey some of the excitement that this exchange of information is generating at the cross-roads of biology, physics, mathematics and engineering.

More concretely, this article contains an introduction to the mathematical study of swimming locomotion of unicellular organisms (e.g., unicellular algae). We give a mathematical formulation of the basic problem of "swimming by prescribing time-histories of shapes", and then use the tools of geometric control theory to identify some general principles governing life at low Reynolds numbers, that can guide the design of engineered devices trying to replicate the successes of their biological counterparts. We also illustrate how these principles are at work in biological organisms, and provide a more detailed case study of the behavior of one specific organism, *Euglena gracilis*, which exhibits a transition from flagellar swimming to amoeboid motion (by propagation of peristaltic waves along the body) in response to increasing confinement.

Besides their primary function, which is motion, locomotion strategies employed by biological organism represent also a rich source of inspiration for studying mechanisms for shape control. They are not visible to the naked eye and revealing them by observing them with a microscope provides a golden mine for new solutions to the problem of controlling shape in order to execute a function. Function often follows from shape in biology. This is true in a wide range of phenomena and across many time-scales, from morphogenesis, to adaptability to changing requirements from an evolving environment, to functional behaviors resting on the possibility of spanning diverse shapes over time, as is the case in motility. We focus on morphing mechanisms based on Gauss' *theorema egregium*, a principle we like to call *Gaussian morphing* [36], and that was pioneered in [67]. Here we can witness in a concrete setting the two-way interaction between biology and mechanics/engineering mentioned above, with shape morphing mechanisms of biological organisms suggesting, for example, new solutions for medical tools for minimally invasive surgery.

We end this introduction by mentioning briefly some topics that are very closely related to the subject of these notes but that, however, are not discussed explicitly here. The conceptual framework we adopt to study swimming locomotion is general, and it applies to other forms of motility besides swimming. Locomotion arises from the mechanical interactions of an active body (i.e., a body capable of changing its shape) with its surroundings, driven by the action–reaction principle. In the case of higher organisms, muscle activity selects a preferred state of deformation, the configuration that the body would acquire in the absence of external forces. Modulating this state of spontaneous deformation in time while in contact with a surrounding medium generates reactive forces from the environment that can be used as propulsive forces for locomotion, just as in the case of a fish waving a fin. Examples of reactive forces exploited for locomotion come from the interaction with a substrate, as in the case of frictional ground forces in human and animal legged locomotion [60], in the limbless undulatory locomotion of snakes [33, 59, 61], and in the peristaltic locomotion of worms [1]. Similarly, hydraulic and (non-newtonian) viscous forces arise from the interaction with a substrate in the case of snails gliding on a substrate [30, 48, 68, 69]. Viscous and inertial forces exerted by the surrounding fluid are the interaction forces with the environment in the case of swimming and flying [32, 72]. Clearly, the list could continue. Anyway, higher organisms with a nervous system, where proprioception and feedback become very important are outside of the scope of these notes. See [60, 62, 86] for some interesting ideas and references in this context.

At the level of single cell locomotion, muscle activity is replaced by activity of molecular motors exerting forces of biofilaments (actin filaments, microtubules) as in the case of the actin cortex of eukaryotic cells. Different modes of motility can arise, such as motion by blebbing, see, e.g., [31, 108], by frictional interaction with a channel or a surrounding fluid arising as a reaction to actin retrograde flow [21], or by lamellipodia protrusion thanks to cycles of attachment at the leading edge (lamellipodium), retrograde actin flow, and detachment at the trailing edge, as in the migration of adhesive cells on or within solid substrates, matrices, and tissues, see [8]. Actin-powered motility of adhesive cells is very common in biology, hence it is a vast topic with a very large literature. Conformational changes of molecular motors, polymerization of actin filaments and, more generally, growth may provide the energy required for motion via biochemical reactions [3, 16, 27, 85, 88, 109].

Modulation of adhesive forces is often crucial for this type of locomotion [94]; in more macroscopic cases, similar stick-slip effects can be obtained through directional friction: see, e.g., [47–49, 54]. Motility of neuronal growth cones has been analyzed in [89]. Contact guidance of adhesive cells by substrate patterning (e.g., chemical guidance with adhesive lines on an otherwise repellent surface, or guidance by curvature with adhesive tubes) is an interesting, related topic, see [25]. Here, the possibility of a statistical mechanics approach based on shape fluctuations of cells is a very attractive recent development. Further discussion on these topics can be found in the work by Lev Truskinovsky and his co-authors, and reported in another section in this volume.

The study of motion in plants [41], e.g. tropisms and nastic movements, is also closely related to the theme of these notes. The fact that the time scales associated with these plant motions are long compared with the typical human attention span does not make them any less interesting, and time-lapse photography can reveal very interesting motile behaviors: see the recent article on nutations of plant shoots [2] and the references quoted therein for an introduction. Complex helical motion motion of growing shoots or root tips is often tied to effective exploration and penetration of the subsoil, or to the search for nearby supports in the case of climbing plants. The oscillatory movements (nutations) of growing plant shoots reveal some striking similarities with the beating of eukaryotic flagella and cilia, although on very different time scales. This is a reflection of the fact that the bio-chemical process that govern the response are very different (bio-chemistry of conformational transformation of the molecular motors in the eukaryotic flagellum case, auxin transport and cell growth in the plant nutation case). However, there seems to be an interesting and yet unexplored connection between the two phenomena, at least at some level, although the details of the response mechanism are certainly very different.

The focus of these notes is on conceptual principles. These are often best extracted from the analysis of model problems arising from simplified minimal systems, that retain the richness of the original problem but can be reduced to simple calculations. As a consequence, we do not talk about numerical techniques to solve control problems associated with locomotion questions, even though this is a very important topic. The reader is referred to [10, 11, 15, 22, 24, 55, 63, 64, 70, 75, 77, 79, 96, 97, 102, 103, 107] and the references quoted therein for examples and analysis of navigation and optimal control problems in biological and bio-inspired locomotion, including numerical strategies for their solution.

1.2 Swimming at Low Reynolds Numbers

We describe the motion of a generic swimmer through a (time-dependent) shape map $t \mapsto \bar{\Phi}_t$, which specifies the way the reference configuration $\mathcal{B}$ evolves in time as seen by an observer moving with the swimmer (we identify this observer with the body-frame), and through the way the body frame moves with respect to the lab-frame. The current position of the body-frame is given by the position of the origin, $\mathbf{c}(t)$, while the orientation of the axes is obtained from the axes of the lab-frame through the rotation $\mathbf{R}(t)$. In formulas (see also Fig. 1.1),

$$\Phi_t(X) = \mathbf{c}(t) + \mathbf{R}(t)\bar{\Phi}_t(X) = (\mathbf{c}(t) + \mathbf{R}(t)id(X)) + \mathbf{R}(t)\bar{\mathbf{u}}_t(X) \qquad (1.1)$$

where, in the second identity, we have written $\bar{\Phi}_t$ as the sum of the identity map id plus a displacement $\bar{\mathbf{u}}_t$. This emphasizes that $\Phi_t(X)$ consists of a rigid motion (the one in brackets), and of a genuine change of shape associated with $\bar{\mathbf{u}}_t$.

The map (1.1) gives the position x at time t of a (material) point $X \in \mathcal{B}$ of the swimmer. Given a point $x \in \mathcal{B}_t = \Phi_t(\mathcal{B})$, this is the position at time t of the point

$$X = \Phi_t^{-1}(x) = \bar{\Phi}_t^{-1}(\mathbf{R}^T(t)(x - \mathbf{c}(t))) \qquad (1.2)$$

The (Lagrangian) velocity of a (material) point of the swimmer is the time derivative of (1.1),

$$\dot{\Phi}_t(X) = \dot{\mathbf{c}}(t) + \dot{\mathbf{R}}(t)\bar{\Phi}_t(X) + \mathbf{R}(t)\dot{\bar{\Phi}}_t(X) \qquad (1.3)$$

Fig. 1.1 Reference and deformed configurations of a swimmer: parametrization of the swimmer motion in terms of position $\mathbf{c}(t)$, orientation $\mathbf{R}(t)$, and shape $\bar{\Phi}_t$. Figure reproduced from [35]

where superposed dots denote time derivatives. The (Eulerian) velocity of the point of the swimmer occupying place x at time t is

$$\dot{\Phi}_t\left(\Phi_t^{-1}(x)\right) = \dot{\mathbf{c}}(t) + \boldsymbol{\omega}(t) \times (x - \mathbf{c}(t)) + \mathbf{R}(t)\dot{\bar{\Phi}}_t\left(\bar{\Phi}_t^{-1}\left(\mathbf{R}^T(t)(x - \mathbf{c}(t))\right)\right) \tag{1.4}$$

where $\boldsymbol{\omega}(t)$ is the axial vector associated with the skew-symmetric matrix $\dot{\mathbf{R}}(t)\mathbf{R}^T(t)$.

Shape changes of the swimmer induce motion of the surrounding fluid. Dealing with microscopic scales (so that the Reynolds number is small) and assuming that the rates at which shape changes occur are not exceedingly fast (so that the Womersley number is also small), we model the flow with the stationary Stokes equations, so that the velocity $\mathbf{u}$ and pressure p in the fluid satisfy

$$\eta \Delta \mathbf{u} - \nabla p = 0 \quad \text{and} \quad \operatorname{div} \mathbf{u} = 0 \quad \text{in } \mathbb{R}^3 \setminus \mathcal{B}_t \tag{1.5}$$

where η is the viscosity of the fluid, together with the no-slip condition at the interface between the fluid and the swimmer boundary

$$\mathbf{u}(x, t)|_{\partial \mathcal{B}_t} = \dot{\Phi}_t\left(\Phi_t^{-1}(x)\right)|_{\partial \mathcal{B}_t} \tag{1.6}$$

and suitable decay conditions at infinity. This outer Stokes problem is well posed, and given the one-parameter family of Dirichlet data $t \mapsto \mathbf{u}(x, t)|_{\partial \mathcal{B}_t}$ (i.e., given the maps $t \mapsto \mathbf{c}(t), \mathbf{R}(t), \bar{\Phi}_t$), the distributions of velocity $\mathbf{u}(x, t)$ and pressure $p(x, t)$ in the fluid are uniquely determined.

The motion of the swimmer is governed by the balance of linear and angular momentum. We neglect inertia, and all other external forces different from those exerted by the fluid. So the balance of linear and angular momentum become the statement that the total force and torque exerted on the swimmer by the surrounding fluid vanish. Denoting the Cauchy stress in the fluid with

$$\mathbf{S}[\mathbf{u}](x, t) = -p(x, t)\mathbf{I} + \eta\left(\nabla \mathbf{u}(x, t) + \nabla \mathbf{u}^T(x, t)\right) \tag{1.7}$$

where $\mathbf{I}$ is the identity, we write these as

$$0 = \mathbf{f}(t) = \int_{\partial \mathcal{B}_t} \mathbf{S}[\mathbf{u}](x, t)\mathbf{n}(x)dA \tag{1.8}$$

and

$$0 = \mathbf{g}(t) = \int_{\partial \mathcal{B}_t} (x - \mathbf{c}(t)) \times \mathbf{S}[\mathbf{u}](x, t)\mathbf{n}(x)dA \tag{1.9}$$

where $\mathbf{n}(x)$ is the outer unit normal at $x \in \partial \mathcal{B}_t$. It turns out that, given $t \mapsto \bar{\Phi}_t$, Eqs. (1.8) determine uniquely the two time-dependent vectors $\bar{\mathbf{v}}(t) = \mathbf{R}^T(t)\dot{\mathbf{c}}(t)$ and $\bar{\omega}(t) = \mathbf{R}^T(t)\omega(t)$, namely, the representations in the body-frame of $\dot{\mathbf{c}}(t)$ and $\omega(t)$. Thus, $\mathbf{c}(t)$ and $\mathbf{R}(t)$ are found by integrating the equations $\dot{\mathbf{c}}(t) = \mathbf{R}(t)\bar{\mathbf{v}}(t)$ and $\dot{\mathbf{R}}(t) = \mathbf{R}(t)[\bar{\omega}(t)]_\times$ (where the skew-symmetric tensor $[\bar{\omega}(t)]_\times$ is defined by $[\bar{\omega}(t)]_\times \mathbf{a} = \bar{\omega}(t) \times \mathbf{a}$ holding for every vector $\mathbf{a}$). This shows that the following

Swimming Problem given a history of swimmer shapes $t \mapsto \bar{\Phi}_t$, find the corresponding history of positions and orientations $t \mapsto \mathbf{c}(t), \mathbf{R}(t)$ has a unique solution. The reader is referred to [40] for a detailed proof of this fundamental fact.

A further interesting question is whether, given initial position and orientation $\mathbf{c}(0), \mathbf{R}(0)$ and a target position $\mathbf{c}(0) + \Delta\mathbf{c}$ (or a target position and orientation pair), there exist a shape history $t \mapsto \bar{\Phi}_t$ such that the swimmer can reach the target. This is a typical question of control theory. In fact, swimming is a perfect example of exploiting fluid-structure interactions to control the (Navier)-Stokes equations: we use shape changes and act on the fluid to produce flows that generate exactly those forces that propel the swimmer in the desired way.

The question of whether a periodic shape change can result in a net displacement $\Delta\mathbf{c}$ in a cycle has attracted a lot of interest in the literature, starting from G.I Taylor's educational movie on low Reynolds number flows [99] and Purcell's seminal paper [87] popularizing some of the seemingly paradoxical aspects of life at low Reynolds numbers. This is the so-called *Scallop Theorem*, stating that, without inertia, a scallop-like organism that can only open and close its (rigid) valves cannot swim. More precisely, a low Reynolds swimmer varying its shape by periodically modulating the opening of its valves can only achieve $\Delta\mathbf{c} = 0$ in one period. In the language of control theory, this is the statement that the state of the system (in particular, its position) is not controllable in terms of the rate of shape change (the input of the system). We will return to this issue in the next section.

1.3 Locomotion Principles and Minimal Swimmers

In this section, we use the framework introduced in Sect. 1.2 to discuss some general locomotion principles and to infer some prescriptions on how to design competent swimmers of minimal complexity. We do this by focusing on some simple, yet representative examples.

1.3.1 *Looping in the Space of Shapes: No Looping? No Party!*

A simple model system which is of great conceptual value is the three-sphere-swimmer proposed in [83]. Consider the case in which $\mathcal{B}$ consists of three rigid spheres of equal radius, whose centers are aligned and constrained to move along

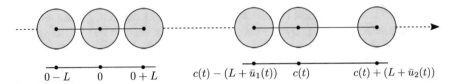

Fig. 1.2 Three-sphere-swimmer: parametrization in terms of position $c(t)$ and shape $(\bar{u}_1(t), \bar{u}_2(t))$. Figure reproduced from [35]

one line parallel to the unit vector $\mathbf{e}$, only varying their mutual distances $L + \bar{u}_1(t)$, $L + \bar{u}_2(t)$ (see Fig. 1.2). The position of every point of the system is specified, once we know the positions of the centers of the three spheres

$$x_1(t) = c(t) - (L + \bar{u}_1(t)), \quad c(t), \quad x_2(t) = c(t) + (L + \bar{u}_2(t)). \quad (1.10)$$

We consider a T-periodic shape change

$$t \mapsto \bar{\mathbf{u}}(t) = (\bar{u}_1(t), \bar{u}_2(t)), \quad (1.11)$$

namely, a closed curve in the space of shapes which we assume to be traced in the anti-clockwise direction.

Linearity of the Stokes system leads to linear dependence of the forces in (1.8) on the Dirichlet data of the outer Stokes problem, hence on $\dot{c}$, $\dot{\bar{u}}_1$, $\dot{\bar{u}}_2$. The component along $\mathbf{e}$ of the force balance is then written as

$$0 = f(t) = f_1(\bar{\mathbf{u}}(t))\dot{\bar{u}}_1(t) + f_2(\bar{\mathbf{u}}(t))\dot{\bar{u}}_2(t) + f_3(\bar{\mathbf{u}}(t))\dot{c}(t) \quad (1.12)$$

where, in view of translational invariance, the force coefficients are independent of position and depend only on shape. It turns out that $f_3 \neq 0$ (in fact, $f_3 < 0$ because it is the drag opposing the motion of the system when the system translates rigidly at unit speed, and the drag has direction opposite to the one of the velocity). Solving for $\dot{c}$ we obtain

$$\dot{c}(t) = \mathbf{V}(\bar{\mathbf{u}}(t)) \cdot \dot{\bar{\mathbf{u}}}(t), \quad \text{where} \quad V_i(\bar{\mathbf{u}}(t)) := -\frac{f_i(\bar{\mathbf{u}}(t))}{f_3(\bar{\mathbf{u}}(t))} \quad (1.13)$$

Using Stokes theorem, we obtain the displacement Δc in one stroke as

$$\Delta c = \int_0^T \mathbf{V}(\bar{\mathbf{u}}(t)) \cdot \dot{\bar{\mathbf{u}}}(t)dt = \int_\omega \text{curl}_{\bar{\mathbf{u}}}\mathbf{V}(u_1, u_2)du_1 du_2 \quad (1.14)$$

where $\text{curl}_{\bar{\mathbf{u}}}\mathbf{V} = \partial V_2/\partial \bar{u}_1 - \partial V_1/\partial \bar{u}_2$ and ω is the region of shape space enclosed by the closed curve (1.11).

As emphasized in [9], formula (1.14) above summarizes several key results of low Reynolds number swimming. The first one is that, if the closed curve $\partial\omega$ spans zero area (i.e., the loop in shape space is trivial, as it happens for a reciprocal shape change), then the displacement vanish. This is the so-called *Scallop Theorem* of [87], already discussed in Sect. 1.2, stating that a scallop-like organism that can only change shape by opening and closing its rigid valves cannot swim in the absence of inertial forces.

The second important result is that, even when the loop in shape space is non-trivial, the displacement is zero if the integrand in (1.14) vanishes. Therefore, swimming rests on the fact that *hydrodynamic resistance forces* (the f_i's defining the vector field $\mathbf{V}$ in (1.13)) *are shape-dependent*, as probed by the differential operator $\mathrm{curl}_{\bar{u}}$.

The third fact following from (1.14) is that the displacement in one stroke is *geometric*: it only depends on the geometry of the loop (1.11) drawn in the space of shapes, not on the speed at which the loop is traced.

Finally, formula (1.14) shows that the system is fully *controllable*. Indeed, if $\Delta c \neq 0$ is the displacement achieved with the loop Γ, smaller displacements of the same sign can be achieved with loops of smaller area, any positive multiple $k\Delta c$ can be achieved by tracing k times the curve Γ, and $-\Delta c$ can be obtained by tracing Γ in the direction opposite to the one used to obtain Δc.

We close this section by remarking that the one-dimensional nature of the swimming dynamics of the three-sphere-swimmer (one degree of freedom for the translational velocity of the swimmer) has made it possible to analyze the problem with simple tools, such as Stokes theorem from differential calculus. In the case of general swimmers (six degrees of freedom for the positional and orientational velocity of the swimmer) the main concept is exactly the same, namely, controllability needs lack of integrability of a differential form. One can obtain similar results in the more general setting by using the tools of Geometric Control Theory. The fact that the curl is non-trivial (i.e., the vector field $\mathbf{V}$ is not the gradient of a scalar potential) is replaced by the requirement that the coefficients of the affine control system governing the dynamics of the swimmer generate, through their Lie-brackets, the whole tangent space. In other (more technical) words, the affine control system governing the dynamics of the swimmer should be bracket-generating, and then controllability is guaranteed by Chow's theorem, see e.g., [9, 13, 80].

1.3.2 Minimal Swimmers With or Without Directional Control

Building on the results of the previous section, we want to ask now the following question. What is the minimal number of independent motors, or controllers, that can allow the three-sphere swimmer to achieve non-zero net displacements? A superficial answer would be that two independent active elements are needed to generate nontrivial loops in the space of shapes. However, as shown in [81], the correct (and, at first sight, surprising) answer is that one active element suffices.

Indeed, by replacing one of the "arms" between two consecutive spheres with a passive spring, and actuating periodically the remaining one, one can still extract net displacements, because the two arm lengths can still describe a loop in the space of shapes. To understand how this arises, consider the two limit cases of very slow and very fast actuation frequencies. At low actuation frequencies, the viscous forces are negligible with respect to the elastic ones, and the system behaves as if it had one rigid arm (hence, no net displacements). At high actuation frequencies, elastic forces are negligible with respect to viscous ones, and the system behaves as a collection of three beads, one of which is free, while the distance between the other two is oscillating. It is relatively straightforward to show that, in this case, long range hydrodynamic interactions lead to synchronization of the three spheres, and the two distances (the arm lengths) oscillate keeping their sum constant. At intermediate actuation frequency ω, elastic and viscous forces compete, and the dynamics of the system leads to the two distances oscillating at the same frequency, but with a frequency-dependent (locked) phase difference which is controlled by the non-dimensional parameter

$$\Omega := \frac{\omega \eta L}{K}. \tag{1.15}$$

Here K is the stiffness of the passive spring, L its rest length, and η is the fluid viscosity. Thanks to this phase lag, the two arm lengths trace a non-trivial loop in the space of shapes, leading to non-zero (frequency dependent) net displacements in one cycle.

The gain in simplicity associated with getting rid of independent control of one the two arms comes with a cost in terms of the performance of the device, namely, loss of controllability. Indeed, since the phase lag is set by system properties that cannot be tuned, see (1.15), the loops in the space of shapes will be traced with a fixed phase lag, hence in a fixed direction. The sign of the displacement is then hard-wired into the system and the three-sphere swimmer with one passive arm can only move with the passive arm ahead, see [81]. Put differently, one motor/controller leads to the minimal system capable of achieving non-zero net displacements, but without a reverse gear. Two independent motors/controllers are necessary for a controllable system.

In closing this section, it is important to emphasize that this simple example shows in a nutshell features that are much more general. Indeed, the dependence of the net displacement in one cycle from the actuation frequency shows that, when shape is not fully prescribed but it rather emerges form the balance between hydrodynamic and elastic forces, the purely geometric picture of Sect. 1.3.1 is no longer sufficient. The behavior of the system arises form a two-way fluid-structure interaction problem, in which one needs to solve for the unknown shape variables by coupling an equilibrium problem for an elastic structure to the dynamics of the surrounding fluid (the outer Stokes problem of Sect. 1.3.1). In this context, deciding on the controllability properties of the swimmer becomes more involved. These controllability questions are very relevant. Examples of these questions are whether a sperm cell can trace curvilinear trajectories (and through which control mechanisms?), whether an artificial sperm-like robot whose flexible magnetic tail is actuated by an external oscillating magnetic field can trace any desired trajectory in space and, in particular, can proceed both forward and backward along the direction of the flagellum when the flagellum experiences small oscillations around a straight extended configuration. Some of these question are further explored below.

1.3.3 Steering by Modulation of the Actuation Speed

Section 1.3.2 has shown that when shape is not completely prescribed, but it rather emerges from the balance of elastic and viscous forces, some adjustments need to be made to the geometric picture of Sect. 1.3.1. The presence of elastically deformable parts in a swimmer makes the net displacement in a cycle frequency-dependent. This fact can be used to steer a swimmer along curved trajectories.

The swimmer studied in [34] consists of a (rigid) spherical head with an elastic tail attached to it, as schematically depicted in Fig. 1.3. We considered planar motions of the system, assuming that the swimmer can actively control only the angle α between the head and the tail. We studied the resulting swimming motion under generic periodic time histories $t \to \alpha(t)$ of the control parameter, resulting in a periodic beating of its elastic tail.

The first surprising feature of the system is the ability of the swimmer to propel itself and to "steer", following either straight or curved trajectories (on average, after many beats), despite being actuated by only *one* control parameter. Secondly, the resulting displacements after each beating period are not geometric: Changing the speed of the periodic control α *does* change the resulting displacement. There is no contradiction, however, between these results and the observations in Sect. 1.3.1. The key to realize this is that, in the swimmer of Fig. 1.3, shape is not completely prescribed, because of the presence of the elastic tail. The shape of the latter is not

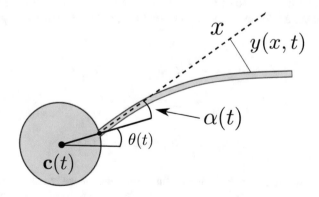

Fig. 1.3 Swimmer with elastic tail. The moving frame of the swimmer is given by the centre **c** and the orientation θ of the head. The swimmer controls the angle α between its spherical head and its tail at the point of attachment. Figure reproduced from [35]

a-priori known, and it only *emerges* from the balance of elastic and hydrodynamic forces arising from the actuation of the angle α. Thus, moving from systems in which shape is completely controlled to systems in which shape is partly emergent, the picture changes completely with respect to the Scallop Theorem scenario of [87].

We considered in [34] the equations of motion of the system in the local drag approximation of Resistive Force Theory, see [70], restricting our analysis to stiff-tailed swimmers. This allowed us to obtain analytical results in the small parameter regime $\epsilon \ll 1$ where ϵ is the ratio between the typical viscous and elastic force acting on the tail (called the Machin number in recognition of the insight contained in his seminal paper [76]). A formula that sheds light on the behavior of the system during its motion is the one proved in [34] for the (normalized) deviation $y(x, t)$ of the tail from its straightened configuration

$$y(x, t) = -\epsilon p(\alpha(t), x)\dot{\alpha}(t) + O(\epsilon^2), \tag{1.16}$$

see Fig. 1.3. The function $p(\alpha, x)$ in Eq. (1.16), which can be calculated explicitly, is a positive polynomial with α-dependent coefficients in the variable x. The dependence of the deviation y on the velocity $\dot{\alpha}$ of the angle α has a simple physical reason: The faster the tail beats, the larger are the viscous forces acting on it, which result in larger bending of the tail itself. The critical consequence of (1.16) follows from another simple observation: given the position of centre $\mathbf{c}$ and the orientation θ of the swimmer's head (see Fig. 1.3), the configuration of the swimmer is fully determined by the angle α and the deviation y. That is, α and y determine the shape of the swimmer. What (1.16) shows, then, is that the shape of the swimmer in motion is fully determined (at least at first approximation) by the angle α and its rate-of-change $\dot{\alpha}$, which, in turn, can be considered as a shape control parameter.

Indeed, a loop in the shape-space of the elastic tail swimmer can be effectively given by the closed curve $t \to (\alpha(t), \dot{\alpha}(t))$. Consistently with the basic principles stated in Sect. 1.3.1, we showed that net displacements $\Delta \mathbf{c}$ and net rotations $\Delta \theta$ of the swimmer arise because of this looping. More precisely, denoting by ω the two dimensional set enclosed by the loop $t \to (\alpha(t), \dot{\alpha}(t))$, we derived the following formulas

$$\Delta \mathbf{c} = \epsilon \int_\omega \mathbf{U}(\alpha, \psi)d\alpha d\psi + O(\epsilon^2) \tag{1.17}$$

and

$$\Delta \theta = \epsilon \int_\omega W(\alpha, \psi)d\alpha d\psi + O(\epsilon^2), \tag{1.18}$$

where $\mathbf{U}$ and W are two non-vanishing functions (vector-valued and scalar-valued, respectively) which are independent on the loop itself. At leading order in ϵ, formulas (1.17)–(1.18) have the exact same structure of Eq. (1.14).

From (1.17)–(1.18) we can deduce the characteristic motion control capabilities of the swimmer. First, we can conclude that propulsion is possible, since a loop $t \rightarrow (\alpha(t), \dot{\alpha}(t))$ naturally spans non-zero area, as one can see in the simple case $\alpha(t) = \sin t$. Second, (1.17)–(1.18) explain why net displacements in one cycle are not geometrical. For example, a simple time rescaling $t \rightarrow \lambda t$ results in a deformation of the shape parameters loop $t \mapsto (\alpha(\lambda t), \lambda \dot{\alpha}(\lambda t))$, thus faster (or slower) beating results in different displacements. More general modulations of the velocity of beating can be considered, resulting in different geometries of the shape parameters loop. This gives room for motion control, so that the swimmer can couple displacements and rotation during each period (steering). In fact, one can show that the swimmer follows curved trajectories when the beating is asymmetric, namely, when for example α has a fast up-beat phase followed by a slow down-beat phase during one shape cycle.

1.3.4 Swimming by Lateral Undulations: Optimality of Traveling Waves of Bending

Swimming by lateral undulation is a very common locomotion strategy that is used by a large number of swimmers both at the macroscopic and the microscopic scale (for the latter, see Sect. 1.4 below for more details). The mechanism by which propagating a traveling wave of bending produces propulsion was analyzed in the seminal paper [98], one of the milestones of the whole literature on biological fluid dynamics at microscopic scales. The traveling waves of bending studied in this paper are of the form

$$\bar{v}(X, t) = b \sin(kX - \omega t) = b \cos(\omega t) \sin(kX) - b \sin(\omega t) \cos(kX), \qquad (1.19)$$

which shows how they represent a non-trivial loop (in fact, a circle, since $\cos^2(\omega t) + \sin^2(\omega t) = 1$) in a space of shapes parametrized by the two wave forms $\sin(kX)$ and $\cos(kX)$. In other words, Taylor's waves of bending fall in the general framework of Sect. 1.3.1.

More in detail, let us consider a planar sheet infinitely extended in the z direction. Let us focus on the plane (x, y), and let X be a Lagrangian coordinate along the axis x, which is assumed to move with the sheet (hence x is the horizontal axis of the body frame). Denoting by $(\bar{u}, \bar{v})$ the components of the displacement of a point $(X, 0)$ in the body frame, we can consider the history of shapes described by a traveling wave of the form

$$\bar{u}(X, t) = a \cos(kX - \omega t - \phi) \qquad (1.20)$$

$$\bar{v}(X, t) = b \sin(kX - \omega t) \qquad (1.21)$$

These functions $\bar{u}$, $\bar{v}$ describe shape waves propagating along the body axis that can be both of stretching and of bending type, with a phase shift ϕ. Waves of stretching ($a > 0$, $b = 0$) can be assimilated to peristaltic waves, waves of bending ($a = 0$, $b > 0$) to lateral undulations. For $\omega > 0$ and $k > 0$, these waves propagate in the direction of increasing x and X.

Computing velocities from positions using (1.20)–(1.21), and using these velocities as Dirichlet boundary conditions for the outer Stokes problem, one can try to solve for the velocities of the fluid and obtain the (horizontal) velocity U of the sheet. This can be done through a series expansion leading to the following expressions for the leading order term in the swimming speed U

$$U = -\frac{1}{2}\omega k \left(b^2 + 2ab \cos\phi - a^2\right) \tag{1.22}$$

The reader is referred to [32] for a proof of these results. In particular, we have

$$U = \frac{1}{2}\omega k a^2, \quad \text{(peristaltic case: wave of stretching)} \tag{1.23}$$

$$U = -\frac{1}{2}\omega k b^2, \quad \text{(undulation case: wave of bending)} \tag{1.24}$$

which show that in the case of a (peristaltic) wave of stretching the motion is prograde (i.e., in the same direction of the direction of propagation of the wave) while, in the (undulatory) case of a wave of bending, the motion is retrograde (i.e., in the direction opposite to the direction of propagation of the wave). Some interesting applications of these formulas to the biological world of snails and earthworms and of bio-inspired robotic replicas can be found in [1, 47–49, 54] and in the references quoted therein.

Moving form the idealized case of the Taylor sheet to the analysis of a concrete swimmer, which is in particular of finite length, is not straightforward. We have studied in a series of papers the case of a planar swimmer consisting of N rigid segments of equal length, connected by rotational joints. For $N = 3$ this is Purcell's three-link minimal swimmer [87], while considering the limit $N \to \infty$ we can reproduce the geometry of Taylor's sheet discussed above. This N-link swimmer has been analyzed in the framework of Geometric Control Theory in several scenarios. One is the case in which the angles between successive links (which provide a discrete analog of the local curvature) are prescribed, i.e., we are dealing with a problem of swimming at prescribed shapes as in Sect. 1.3.1, see [12]. It turns out that the governing equations for this swimmer have the structure of an affine control system without drift, for which one can apply powerful theorems to prove controllability. Another case is obtained by assuming that the links are ferromagnetic and a time-dependent (oscillatory) external magnetic field is applied, and that elastic torsional springs are present at the joints between successive links. The resulting N-link magnetic swimmer, analyzed in [14], provides an example in which shape emerges only from the balance of elastic restoring torques, external magnetic torques, and torques arising from hydrodynamic drag, i.e., a problem conceptually analogous to the one of Sect. 1.3.2. The governing equations for this second case have the structure of an affine control system with drift, for which general sufficient conditions for controllability are not available. Loss of controllability for swimmers whose shape is only partially controlled can therefore be understood in the light of the structure of the governing equations, using the tools of Geometric Control Theory.

The N-link swimmer can also be used to probe questions of optimal control. Deformations in the form of waves traveling along the body are very common in nature (both for swimmers and crawlers), and are used very frequently in slender bio-inspired mobile robots. Are they optimal in any sense, when compared to alternative actuation strategies? We have examined this question in [15]. More precisely, we have considered the problem of finding the gait of minimal energy expenditure among all those capable of reaching a prescribed displacement in one shape cycle.

This problem of optimal control is nonlinear in the shape parameters (the angles between successive joints) and finding (even numerically, when N is large) the optimal gait explicitly is exceedingly difficult. By considering the regime of small deviations from the straight configuration, i.e., under the assumption of small-amplitude angles, and considering the approximation of the governing equations at leading order in the shape parameters, we obtain an affine control system that can be analyzed in full detail. We find that optimal gaits are always two-dimensional elliptical loops, independent of N. These gaits bridge Purcell's loops for the two-dimensional shape space associated with $N = 3$, to gaits that, modulo edge effects, can be identified with Taylor's traveling waves of bending for large N, of a type similar to (1.19).

The result above, namely, the energetic optimality of undulation waves as a swimming strategy for a swimmer differing only slightly from a straight segment only depends on structural properties and symmetries of the governing equations, which in turn reflect the geometric symmetry of the physical problem at hand. In fact, in this regime of small-amplitude joint angles, the perturbations from the rectilinear geometry of the reference configuration are small, and a slender one-dimensional swimmer with homogenous geometric and mechanical properties that interacts with a homogeneous surrounding medium is a system which is essentially invariant under shifts along the body axis. This is exactly true for an infinite or a periodic system and approximately true, modulo edge effects, for a system of finite length. The relevance of traveling waves as optimal gaits is therefore naturally suggested by the geometric symmetry of the system and, in fact, it emerges naturally from the symmetries of the governing equations. Given the generality of this argument, it is quite natural that the same conclusion holds true for other types of locomotion as well, such as the case of one-dimensional crawlers gliding on solid surfaces, see [1].

Removing the assumption of small-amplitude joint angle and exploring the case in which large deviations of the shape from the rectilinear one are allowed is difficult. Numerical simulations show that the optimal gaits (obtained for the case

$N = 3$ and $N = 5$ in [15]) are planar but non-convex in the case $N = 3$, and non-planar with complex geometries in the case $N = 5$. This is not surprising because, when the restricted setting of small perturbations from the rectilinear geometry is abandoned, and large shape changes are considered, then invariance under shifts along the body axis is lost and traveling waves are no longer a natural basis for the study of the properties of optimal gaits. It would be interesting to find out whether the closed curves of high dimensionality representing the optimal gaits in the large deformation regime, that we find numerically, do exhibit special structural properties or symmetries: this question is, at present, completely open.

1.4 Biological Swimmers

It may seem surprising that the principles discussed above may be relevant in the study of the locomotion strategies of unicellular organisms, whose bodies look very different from the idealized systems of beads and springs analyzed in the previous section. In fact, this is indeed the case, as we argue in the remainder of this section. We will show that the principles discussed in Sect. 1.3 go a long way in rationalizing behaviors observed in biology.

1.4.1 Chlamydomonas' Breaststroke

Chlamydomonas is a unicellular alga, with a round body, which swims thanks to the beating of two flexible anterior flagella, see e.g. [56, 58]. It has been used as a model organism, in particular for what concerns flagellar locomotion [51].

In one of its typical behaviors, the cell beats the two flagella in synchrony in a plane, symmetrically about a central symmetry axis of the body. With this perfect breast-stroke, the cell progresses with a rocking back-and-forth motion along the symmetry axis. Net displacements are made possible by the fact that the shape of the flagella varies during one stroke cycle. They are extended during the power phase of the stroke, when the flagella beat (say) downwards, and push the cell body upwards. They are contracted in the recovery part of the stroke, when they move upwards to recover the initial posture, and push the cell downwards (hence the rocking motion). Net upward displacement results from the fact that the hydrodynamic forces generated by the extended flagellum are higher than those generated by the retracted flagellum, moving in the opposite direction. Considering as shape variables the angle formed at the attachment of the body (localized curvature), and a measure of global curvature, we recognize that Chlamydomonas' breast stroke consists of a loop in the space of shapes.

This mechanism, based on the non-reciprocal time-periodic beating of slender one-dimensional structures, is ubiquitous also in the self-propulsion of ciliates, which are typically covered by numerous arrays of beating cilia. Their beating is organized in periodic spatio-temporal patterns called metachronal waves. Individual cilia oscillate back and forth with a shape asymmetry between an extended configuration in the power phase of the stroke and a more bent one during the recovery phase of the stroke.

1.4.2   Sperm Cells and Flagellar Beat

Sperm cells are among the most thoroughly studied examples of unicellular swimmers, see e.g. [53]. There is hardly a more evident illustration of how cell motility is relevant to life as the highly oscillatory motion by which a sperm cell successfully swims its way until it reaches and fecundates an egg cell.

Sperm cells move by beating a flagellum, whose structure is highly conserved across all eukaryotic species. It consists of longitudinal bundles of microtubules, arranged in a precise spatial structure (the $9+2$ architecture of microtubule doublets in the axoneme), on which molecular motors (dyneins) exert forces that cause the creation and propagation of longitudinal bending waves [8]. These bending waves generate the propulsive forces powering the motion of the cell. The same flagellar architecture and the same flagellar beat, with the resulting force-exchange with a surrounding fluid are at the basis of the locomotion strategies of all flagellates and ciliates. In addition, they are at the root of some fluid transport phenomena of great physiological relevance in humans and animals (such as muco-ciliary clearance, i.e, the self-clearing mechanism of the airways in the respiratory system, but cilia may also be involved in the flow of cerebrospinal fluid) which are driven by the beating of ciliated cells lining the walls of the organs inside which the flow takes place.

Interestingly, a basic and fundamental model for the flagellar beat in eukaryotic cilia and flagella is still lacking, in spite this being one of the most thoroughly studied topics in cell motility for the last several decades. The seminal paper by Machin [76] established that the observed wave patterns are incompatible with the hypothesis that the flagellum is a passive elastic filament set in motion by external actuation (e.g., an active process located at the proximal end, as is the case for bacterial flagella, which are driven by a rotary motor located at the proximal end of the flagellum). For eukaryotic flagella, active contractile elements must exist along the length of the flagellum. In other words, mechanics shows that the observed bending waves require the presence of internal actuation along the flagellum: distributed active forces/torques along the flagellum, which we now know to be the result of the action of molecular motors (dyneins) on microtubule bundles But how is the active beat generated? How is the frequency of the beat set? Is it affected by environmental cues and, if so, through which mechanisms? Already at its most basic level, a bending wave requires spatio-temporal patterns of curvature of variable sign along the body axis. This has led to the hypothesis there must exist some coordination mechanism between molecular motors acting on opposite sides of the flagellum: In order to generate the flagellar beat, molecular motors must switch on and off, respectively, at the opposite sides of the bending plane (this is called the 'switch-point hypothesis' in the biophysical literature). Whether this hypothesis is correct and, if so, what controls the switching of activation of axonemal motors is still a debated issue, and it is the subject of active ongoing research [74]. Recent advances in high resolution microscopy (cryo-electron tomography) have enabled us to observe distinctive asymmetric spatial patterns of active and inhibited motors along the length of a flagellum [73]. To date, however, there are no time-resolved observations of the beat pattern, allowing for a direct visualization of the configurations of motors and microtubules during the beat. In addition, it is not known how the beat is regulated, and it is not ruled out that oscillations simply emerge spontaneously as a resonance of the axoneme, seen as mechanical system in which motors respond to the forces transmitted by the microtubules [92]. The quest for a fundamental model of the regulation mechanism of the flagellar beat is still outstanding.

1.5 Euglena Gracilis: A Case Study in Biophysics and a Journey from Biology to Technology

We have been studying Euglenids, and *Euglena gracilis* in particular (see [71]), already for some years. The reason behind our interest in this protist, a unicellular flagellate, is that it exhibits two distinct forms of motility. One is through the beating of a single anterior flagellum (swimming motility). Another one is through very large, elegantly coordinated, rhythmic shape changes of the whole body (amoeboid motion or metaboly). What controls the switching between this two distinct 'gaits' is an interesting question, which is still open, and on which we have some hypotheses.

1.5.1 Metaboly and Mechanisms for Shape Change, Embodied Intelligence

The large shape changes associated with metaboly rely on the special body architecture of *Euglena* cells whose outer envelope, just like other Euglenids, has a complex structure. Indeed, underneath the plasma membrane, they have an ultra-structure called *pellicle*, see Fig. 1.4. This is a complex made of protinaceous strips, microtubules, and molecular motors. The strips have overlap regions and are able to slide one on another along their length. The sliding is powered by molecular motors that induce sliding in the microtubules that run parallel to the strips, along the overlap region. The large shape changes of metaboly correlate closely with geometric rearrangements of the pellicle structure. In fact, the relative sliding along the strips can be thought of as a mechanism of active surface shearing or, in the language of differential geometry, of active change in the surface metric. In view of Gauss' Theorema Egregium, modulating the pellicle shears means modulating

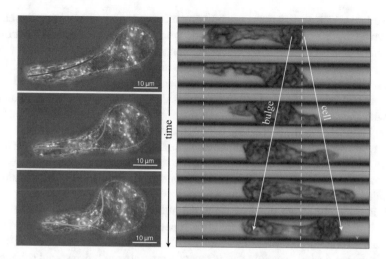

Fig. 1.4 Left: a sample of *Euglena gracilis* imaged in bright-field, reflected light microscopy while exhibiting cell body shape changes (metaboly) concomitant with the reconfiguration of the striated cell envelope. Right: micrographs of *Euglena gracilis* effectively crawling in a capillary under significant spatial confinement by means of peristaltic shape changes. Notice the forward motion of the leading edge of the cell and the corresponding retrograde motion of the traveling bulge sliding against the capillary wall. Figure reproduced from [3]

the surface metric along the surface, and this can produce (Gaussian) curvature. In particular, the propulsive mechanism associated with metaboly consists of the propagation of a round protruding bulge along the axis of an elongated body of approximately cylindrical shape. A surface with a bulge is one with nonzero Gaussian curvature, so that metaboly relies on the propagation of nonzero Gaussian curvature along the cell body. This can be accomplished and, as experiments show, is in fact accomplished by modulation of the pellicle shears along their lengths. And in the regions where the bulge forms, the pellicle strips acquire a characteristic helical shape. This mechanism is described in detail in [17, 18], and in Sect. 1.6 below.

Observation at the microscope of the behavior of *E. gracilis* in capillaries of decreasing diameter or in environments of increasing crowding suggest that metaboly may be triggered by confinement. In fact, we could confirm this hypothesis in [85] by examining swimming *Euglena* cells in environments of controlled crowding and geometry, see Fig. 1.4. Under these conditions of increasing confinement, metaboly allows cells to switch from unviable flagellar swimming to a new and highly robust mode of fast crawling, which can deal with variable levels of geometric confinement, even extreme ones, and turn both frictional and hydraulic resistance into propulsive forces.

To understand how a single cell can control such an adaptable and robust mode of locomotion, we developed in [85] a computational model of the motile apparatus of *Euglena* cells consisting of an active striated cell envelope. The activity of the motors results in tangential forces applied to the pellicle strips along their region of overlap, which causes them to slide one relative to the other. Using the energy consumed by these motors as the input, we assign spatio-temporal patterns of activation generating traveling waves of tangential forces and, in turn, traveling waves of sliding displacements which produce peristaltic waves of the cell body. One of the most striking conclusions that can be reached with the model is the following. Driving the active cell envelope inside capillaries of decreasing diameter, but with the *same* spatio-temporal patterns of activation, the system is capable of quickly adjusting to the increasing confinement, always finding an effective gait (a limit cycle) up to the most extreme levels of confinement. In other words, our model shows that gait adaptability does not require specific mechano-sensitive feedback but can be explained instead by the mechanical self-regulation of an elastic and extended motor system.

From an engineering point of view, the ability of the pellicle to mechanically self-adapt and maintain the locomotory function under different geometric and mechanical conditions represents a remarkable instance of mechanical or embodied intelligence, a design principle that is recently emerging in bio-inspired robotics by which part of the burden involved in controlling complex behaviors is outsourced to the mechanical compliance of the materials and mechanisms that build the device.

In conclusion, our analysis identifies a locomotory function and the operating principles of the adaptable peristaltic body deformation of *Euglena* cells. Further details on metaboly as a form of motility, and on the mechanisms controlling how *Euglena* switches under confinement from flagellar swimming to this mode of behavior are discussed in [85].

1.5.2 Flagellar Swimming, Helical Trajectories and a Principle for Self-Assembly

When left undisturbed in free space, *Euglena* cells swim by beating a single anterior flagellum. Contrary to what happens for Chlamydomonas, see Sect. 1.4.1, the flagellar shapes of *Euglena* are typically non-planar (their geometry is often referred to as 'figure eight' or 'spinning lasso'). The resulting trajectories are also non-planar: *Euglena* cells swim with a characteristic swinging motion (*Erschuetterung*) with apparent sinusoidal trajectories when the cells are imaged in the two-dimensional world of optical microscopy. In fact, as it turns out, this is the typical footprint of a helical swimming motion projected on a 2D plane.

This is in fact proved in [90], where we have managed to reconstruct the three-dimensional trajectories and flagellar shapes of swimming Euglena, starting from time-sequences of two-dimensional images obtained with an optical microscope. This has been made possible thanks to the precise characterization of the orbits (the maps $t \mapsto \mathbf{c}(t), \mathbf{R}(t)$ of the swimming problem of Sect. 1.2) traced by an object propelled by a flagellum beating periodically in time. Hydrodynamics of Stokes flows dictates the following *universal law of periodic flagellar (and ciliar) propulsion*: the orbits of any object propelled by periodically beating flagella (or cilia), swimming far away from walls, boundaries, etc. are always generalized helices. This makes the reconstruction of three-dimensional trajectories possible: once the three-dimensional geometric structure of the orbits is known, these can be recovered from their two-dimensional projections, hence *lifting* the two-dimensional experimental images to three dimensions.

In fact, the same result on helical trajectories holds true if the organism is propelled by an array of cilia, beating periodically. This situation is typically modeled treating the swimmer as a *squirmer*, i.e. a rigid body with a distribution of slip velocities on its boundary, which represent the relative velocity of the fluid (which moves like the tips of the cilia) with respect to the base of the cilia, attached to the boundary of the body. If the slip velocity field is periodic in time, the resulting orbits are generalized helices. For the sake of simplicity, we will illustrate the results on helical swimming above by only considering the discrete curve traced by an arbitrary point of the swimmer (the origin $c(t)$ of the body-frame), following its positions after integer multiples of the beating period T. We have the following:

Helix Theorem The discrete trajectory $k \mapsto c_k = c(0 + kT)$ traced by a micro-swimmer moving in free-space, propelled by the T-periodic beating of a flagellum or of an array of cilia, is a discrete circular helix.

This result is a special case of a more general one proved in [90], but we give here a direct proof following [35]. Assume that, at $t = 0$, body-frame and lab-frame coincide: $c(0) = o, R(0) = I$, and let d and R be the displacement and rotation at the end of one beat cycle. After each cycle, the incremental displacement and rotation *in the body-frame* will always be d and R since the shape change cycle is the same, and the swimming problem (written in the body-frame) is invariant by rotation and translation. Composing these (constant) translations and rotations with the motion of the body-frame, the discrete trajectory in the lab-frame will be

$$c_k = o + d + Rd + R^2d + \ldots + R^{(k-1)}d, \quad \text{and} \quad R_k = R^k, \tag{1.25}$$

where $R_k = R(0 + kT)$ and $R^0 = R(0) = I$.

To see that $(1.25)_1$ is a discrete circular helix, we will use the discrete version of a well known result form differential geometry, stating that a curve with constant curvature and torsion is necessarily a circular helix. This is easy to prove by integrating Frenet's formulas, and is sometimes referred to as Lancret's theorem. To compute the (discrete) curvature at point c_k, consider the (osculatory) plane generated by the three points (c_{k-1}, c_k, c_{k+1}) and spanned by the (discrete) tangents $R^{(k-1)}d$ and $R^{(k-1)}Rd$, see Fig. 1.5. By Frenet's formulas, the curvature at c_k is $1/|d|$ times the angle θ between d and Rd, which is independent of k. Similarly, to compute the torsion at point c_k, consider the binormal at c_k. This is orthogonal to the osculatory plane at c_k, hence parallel to $R^{(k-1)}(d \times Rd)$. Similarly, the binormal at c_{k+1} is orthogonal to the plane spanned by R^kd and $R^{(k+1)}d$, hence parallel to $R^{(k-1)}(R(d \times Rd))$. By Frenet's formulas, the torsion at c_k is $1/|d|$ times the angle ϕ between $d \times Rd$ and $R(d \times Rd)$, which is independent of k. Special cases are

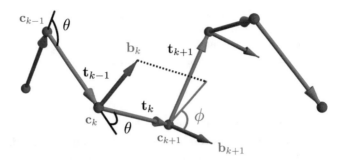

Fig. 1.5 Tangents and binormals to the discrete helix to compute discrete curvature $\theta/|\mathbf{d}|$ and discrete torsion $\phi/|\mathbf{d}|$. Figure reproduced from [35]

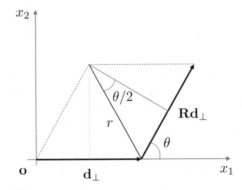

Fig. 1.6 Geometric construction of the parameters of the discrete circular helix associated with shift $\mathbf{d}$ and rotation $\mathbf{R} = \mathbf{R}_\mathbf{e}^\theta$ of angle θ and axis $\mathbf{e}$. In the figure, $\mathbf{d}_\perp$ denotes the projection of $\mathbf{d}$ on the plane perpendicular to the rotation axis $\mathbf{e}$. Figure reproduced from [35]

that of a straight trajectory, which arises when $\mathbf{d}$ is parallel to the axis of $\mathbf{R}$ (hence $\mathbf{d} \times \mathbf{Rd} = 0$, zero discrete curvature), and of a circular trajectory, which arises when $\mathbf{d}$ is perpendicular to the axis of $\mathbf{R}$ (so that both $\mathbf{d} \times \mathbf{Rd}$ and $\mathbf{R} \times (\mathbf{d} \times \mathbf{Rd})$ are parallel to the axis of $\mathbf{R}$ and hence, zero discrete torsion).

The informal discussion above can be made rigorous by using the results in [28], that show that curve $(1.25)_1$ can be seen as the discretization of a continuous curve having as curvature and torsion exactly the values computed from $(1.25)_1$.

It is interesting to give a concrete geometric representation of the helix above, see Fig. 1.6. Let $\mathbf{e}$ be the axis of the rotation $\mathbf{R}$ (the eigenvector corresponding to its eigenvalue equal to $+1$) and θ the angle. We can highlight the two parameters characterising the rotation $\mathbf{R}$ by writing $\mathbf{R} = \mathbf{R}_\mathbf{e}^\theta$. In a reference frame with origin $\mathbf{o}$, first axis aligned with $\mathbf{d}_\perp$, the projection of $\mathbf{d}$ on the plane perpendicular to the rotation axis $\mathbf{e}$ (we are assuming here that $\mathbf{d}$ is not parallel to $\mathbf{e}$, for otherwise the trajectory is a straight line parallel to $\mathbf{d}$), and third axis aligned with $\mathbf{e}$, the equation of the discrete circular helix $(1.25)_1$ is

$$x_1(k) = r \cos\left(\frac{\pi - \theta}{2}\right) + r \cos\left(-\frac{\pi + \theta}{2} + k\theta\right) \tag{1.26}$$

$$x_2(k) = r \sin\left(\frac{\pi - \theta}{2}\right) + r \sin\left(-\frac{\pi + \theta}{2} + k\theta\right) \tag{1.27}$$

$$x_3(k) = 0 + k |\mathbf{d}_{//}| \tag{1.28}$$

where

$$r = \frac{|\mathbf{d}_\perp|/2}{\sin(\theta/2)}, \quad \mathbf{d}_{//} = (\mathbf{e} \otimes \mathbf{e})\,\mathbf{d}, \quad \mathbf{d}_\perp = (\mathbf{I} - \mathbf{e} \otimes \mathbf{e})\,\mathbf{d}. \tag{1.29}$$

Finally, it is interesting to notice that, since the axis of body rotation $\mathbf{R}$ is also the screw axis of the helix traced by the cell body, as the cell moves in average along the screw axis, the body rotates so that the lateral surfaces containing the eyespot (in which a light receptor is located) is periodically exposed to or shaded from light, unless the screw axis is aligned with the light source. In fact, this is a vivid example of a general biochemical mechanism, repeatedly found in nature, whereby periodic signals are used as a tool for navigation: the existence of a periodic signal implies lack of alignment [56]. Thus, the organism can react to the signal by perturbing its beat until the periodic signal is suppressed, in the aligned state. In this way, the coupling between direction of average motion and axis of body rotations could explain the navigation mechanism used by phototactic Euglena to orient with a light source.

The result contained in the Helix Theorem is remarkable in its universality, and it has far reaching consequences for the swimming of flagellated and ciliated unicellular organism: They all trace helical trajectories. This is at least true over time windows over which they exhibit stereotyped behavior (e.g., no turns), resulting on periodic beating of their cilia and flagella. As a matter of fact, the recent biophysical literature abounds with reports of discoveries of the helical structure of experimentally observed trajectories of a variety of micro-swimmers of great biological relevance (bacteria, sperm cells, etc.). Most of the previous literature had focused on the special cases of straight and circular trajectories, which are easily measured from two-dimensional images from an optical microscope, while the reconstruction of trajectories such as helices, which are spatial curves, requires that we resolve the third dimension perpendicular to the focal plane of the microscope. This entails technical problems, and only recently is it becoming possible to overcome them.

The interest and possible applications of the Helix Theorem above are not confined to trajectories of biological (and, possibly, artificial) micro-swimmers. In fact, the theorem yields a principle for the self-assembly of helical structures. Indeed, the theorem characterizes the geometry of any chain resulting from the assembly of rigid monomers, when the position and orientation of the $(k + 1)$-th monomer are constrained to be those arising from a shift $\mathbf{d}$ and a rotation $\mathbf{R}$ of the k-th one. In this case, a discrete circular helix will emerge. In other words, if two monomers can only bind in a precise relative position and orientation, then a self-assembled helical structure will be the outcome when they form in a chain. This simple observation provides a rationale explaining why helices are so ubiquitous in biology and nature, and why bio-polymers self-assemble into helices. For the same reason, self-assembled helical structures are frequently encountered in polymer science, chemical engineering and materials science.

1.6 Shape Control and Gaussian Morphing

The mechanism by which *Euglena* changes shape when executing metaboly, discussed in Sect. 1.5.1, is based on active shears of its outer envelope, which arise in turn from the sliding of its pellicle strips. To the best of our knowledge, this mechanism has never been used in engineered devices, at least until now. For this reason, it is interesting to place this mechanism within a more general discussion of shape control in biological organism, of deployable structures, of bio-inspired morphing structures, of shape programming of active materials, adaptable structures, and the various other variants of these concepts.

In fact, the active shearing explaining *Euglena*'s metaboly is but one example of shape control of two-dimensional thin, shell-like objects, whereby shape control is enforced through control of curvature. There are two main avenues to achieve this. The first one operates by inducing differential strains along the thickness, as in Timoshenko's bimetallic strips [100] and in modern soft variants, see e.g. [5, 29, 38, 44, 93]. The second one operates by controlling in-plane stretches and exploiting Gauss' *theorema egregium*. This second avenue, based on the fact that Gaussian curvature is associated with derivatives of the components of the metric tensor, i.e., differential stretches of the mid-surface, has received a lot of attention in the recent literature, see e.g., [67, 78, 82, 91] and many others. We call *Gaussian morphing* this second strategy, namely, the idea of controlling curvature (shape) of a thin two-dimensional structure through modulated stretching of the mid-surface (via prescription of the metric tensor). The mechanism used by *Euglena* is of this second type.

Many results are available on how this Gaussian morphing principle is at work in biological structures [6, 17, 18], and on how it can be exploited in artificial structures by using for example hydrogels or nematic elastomers, see e.g. [7, 65, 95, 106]. But, up to now, technology has mostly relied on the first principle (differential strains across the thickness), rather than on the second (differential stretches of the mid-surface). Understanding how Gaussian morphing works in specific examples may help popularize this new approach and inspire novel applications in the context of deployable or adaptive structures and devices. With these motivations in mind, we explore two concrete examples in the next sections. Further details can be found in [36].

1.6.1   Controlling the Shape of Surfaces by Prescribing Their Metric

We start by considering the reference configuration of a material surface, namely, a two-dimensional surface immersed in $\mathbb{R}^3$ and its deformations. This means that we consider a map $(u, v) \mapsto \chi_0(u, v) \in \mathbb{R}^3$, where $(u, v) \in (L_0, H_0) \subset \mathbb{R}^2$. A deformed configuration of this material surface will be given by another map, say, $(u, v) \mapsto \chi(u, v) \in \mathbb{R}^3$, again with $(u, v) \in (L_0, H_0) \subset \mathbb{R}^2$.

By computing the surface deformation gradient F and the right Cauchy-Green strain $C = F^T F$, we obtain the metric tensors of the material surface in the reference and deformed configurations as

$$C_0 = g_0 = \begin{bmatrix} \chi_{0,u} \cdot \chi_{0,u} & \chi_{0,u} \cdot \chi_{0,v} \\ \chi_{0,u} \cdot \chi_{0,v} & \chi_{0,v} \cdot \chi_{0,v} \end{bmatrix} \tag{1.30}$$

and

$$C = g = \begin{bmatrix} \chi_{,u} \cdot \chi_{,u} & \chi_{,u} \cdot \chi_{,v} \\ \chi_{,u} \cdot \chi_{,v} & \chi_{,v} \cdot \chi_{,v} \end{bmatrix} = \begin{bmatrix} E & F \\ F & G \end{bmatrix}, \tag{1.31}$$

where a comma denotes partial differentiation.

We are interested in inducing controlled changes of the shape of the material surface by generating changes of lengths and angles of its material fibers through actuation, described by changes of the metric tensor from its reference value g_0 to a new value g. The possibility of changing curvature (morphing) of a surface by acting on its metric is recognized by a remarkable theorem by Gauss, his celebrated *theorema egregium*, stating that the Gaussian curvature K of a surface (the product of its principal curvatures) can be computed by differentiating the components of its metric tensor as

$$- EK = (\Gamma_{12}^2)_{,u} - (\Gamma_{11}^2)_{,v} + \Gamma_{12}^1 \Gamma_{11}^2 + \Gamma_{12}^2 \Gamma_{12}^2 - \Gamma_{11}^2 \Gamma_{22}^2 - \Gamma_{11}^1 \Gamma_{12}^2, \tag{1.32}$$

where $\Gamma_{\beta\gamma}^\alpha$, $\alpha, \beta, \gamma = 1, 2$, are the Christoffel symbols, see [50]. The interpretation of Gauss' *theorema egregium* as a morphing scheme has been pioneered in the seminal paper [67].

We will discuss examples that are motivated by shape changes exhibited by unicellular organisms, of special relevance in the study of cell motility. Our first example, motivated by *Euglena*'s metaboly introduced in Sect. 1.5.1, is the local simple shear arising from the sliding of pellicle strips making up the cell envelope of euglenids. This mechanism for shape change has been analyzed in great detail in [17, 18, 84, 85], and it is described by a metric tensor of the form

$$g = \begin{bmatrix} 1 + \gamma^2 & \gamma \\ \gamma & 1 \end{bmatrix} \quad \text{(shearing mechanism)}. \tag{1.33}$$

Here $\gamma = \gamma(u, v) \in \mathbb{R}$ is the local simple shear between material fibers aligned with the coordinate lines (the direction of the centerline of the pellicle strips in the case of euglenids), an area preserving deformation. Actually, the same mechanism drives the flagellar and ciliary beating in eukaryotic cells discussed in Sect. 1.4.2, where molecular motors induce relative sliding between parallel bundles of microtubules lying on the outer cylindrical envelope of the flagellum, see [36] for more details. Substituting (1.33) into (1.32) we obtain

$$K = \left(\gamma_{,u} - \gamma \gamma_{,v} \right)_{,v} \qquad \text{(shearing mechanism)}. \qquad (1.34)$$

Our second example is motivated by the observations in [39] of the deformations of *Lacrymaria olor*, a uniceliuar ciliate that is easily found in freshwater ponds, just as *Euglena*. This same mechanism is also at work in the braided sheaths of pneumatic artificial muscles of McKibben type [101]. It consists of a stretch with principal directions along the coordinate lines

$$g = \begin{bmatrix} \lambda^2 & 0 \\ 0 & \mu^2 \end{bmatrix} \qquad \text{(stretching mechanism)}, \qquad (1.35)$$

where $\lambda = \lambda(u, v) \in (0, +\infty)$ and $\mu = \mu(u, v) \in (0, +\infty)$ are the stretches along the $u-$ and $v-$coordinate lines, respectively. These are typically the diagonals in the rhombus-shaped unit cell of a meshwork which deforms as a pantograph in the sheath of a McKibben actuator or, for the biological case, in the arrays of biofilaments making up the cell envelope.

The deformation associated with (1.35) is area preserving if $\lambda\mu = 1$, in which case it is called a pure shear. Substituting (1.35) into (1.32) we obtain

$$K = -\frac{1}{\lambda\mu} \left(\left(\frac{\lambda_{,v}}{\mu} \right)_{,v} + \left(\frac{\mu_{,u}}{\lambda} \right)_{,u} \right) \qquad \text{(stretching mechanism)} \qquad (1.36)$$

in the general case while, in the area-preserving case, we have

$$K = - \left(\lambda\lambda_{,v} \right)_{,v} - \left(\mu\mu_{,u} \right)_{,u} \qquad \text{(stretching mechanism}, \ \lambda\mu = 1). \qquad (1.37)$$

1.6.2 Axisymmetric Surfaces

For simplicity, we focus now on axisymmetric shape-shifting surfaces. As reference configuration S_0, we consider the cylinder of radius R_0 such that

$$\chi_0(u, v) = \left\{ R_0 \cos\left(\frac{u}{R_0} \right), \ R_0 \sin\left(\frac{u}{R_0} \right), \ v \right\}, \quad u \in (0, L_0), v \in (0, H_0),$$
$$(1.38)$$

where $L_0 = 2\pi R_0$. This has the identity matrix as metric tensor g_0 and zero Gaussian curvature $K_0 = 0$, in agreement with formulas for g and K in the previous section obtained by setting either $\gamma = 0$ or $\lambda = \mu = 1$.

We are then interested in deformed configurations with axisymmetric shape S, which can be written by assigning a generating curve $\{r(v), z(v)\}$ in the symmetry plane and an azimuthal displacement $\psi(v)$, leading to

$$\chi(u, v) = \left\{ r(v) \cos\left(\frac{u}{R_0} + \psi(v)\right), r(v) \sin\left(\frac{u}{R_0} + \psi(v)\right), z(v) \right\}, \quad (1.39)$$

$$u \in (0, L_0), \, v \in (0, H_0).$$

Substituting (1.39) into (1.31) we obtain

$$\begin{bmatrix} (r/R_0)^2 & r^2\psi'/R_0 \\ r^2\psi'/R_0 & r'^2 + z'^2 + r^2\psi'^2 \end{bmatrix} = \begin{bmatrix} E & F \\ F & G \end{bmatrix} = g, \quad (1.40)$$

where a prime $(\cdot)'$ denotes differentiation with respect to v. Clearly, since the left hand side in the last equation depends only on v, only metric tensors $g = g(v)$ that are independent of u (axi-symmetric actuation) are compatible with (1.39). In these circumstances, Eqs. (1.34), (1.36), and (1.37) from the last section simplify to

$$K = -\left(\gamma\gamma_{,v}\right)_{,v} \quad \text{(shearing mechanism)}, \quad (1.41)$$

$$K = \frac{1}{\lambda\mu}\left(\frac{1}{\mu^2}\lambda_{,v}\mu_{,v} - \frac{1}{\mu}\lambda_{,vv}\right) \quad \text{(stretching mechanism)}, \quad (1.42)$$

$$K = -\left(\lambda\lambda_{,v}\right)_{,v} \quad \text{(stretching mechanism, } \lambda\mu = 1\text{)}, \quad (1.43)$$

respectively.

We would like to compute the axisymmetric shapes that can result from axisymmetric actuation patterns either in simple shear, $\gamma = \gamma(v)$, or in pure shear $\lambda = \lambda(v), \lambda\mu = 1$. From Eq. (1.40) we immediately see that, since $E = (r/R_0)^2$, whenever the metric g is constant, then the axisymmetric surface χ is a cylinder of radius $r = E^{1/2}R_0$, a special instance of a surface with zero Gaussian curvature $K = 0$. This case of constant metric g is the simplest to examine, and we shall consider this case first.

1.6.3 Cylinders from Cylinders

When $K = K_0 = 0$, the axisymmetric morphing surface can be developed onto a plane both before and after actuation. Following [17], in order to study the shape change $S_0 \mapsto S$ induced by the change of metric $g_0 \mapsto g$, it is useful to analyze the process by first cutting S_0 along a direction parallel to the cylinder axis and unfolding it (isometrically) to a plane, then deform this plane with a two-dimensional affine map $\Phi(u, v)$ inducing the (spatially uniform) change of metric $g_0 \mapsto g$

$$\begin{bmatrix} \Phi_{,u} \cdot \Phi_{,u} & \Phi_{,u} \cdot \Phi_{,v} \\ \Phi_{,u} \cdot \Phi_{,v} & \Phi_{,v} \cdot \Phi_{,v} \end{bmatrix} = g = \begin{bmatrix} E & F \\ F & G \end{bmatrix}, \tag{1.44}$$

and then roll-up (isometrically) the deformed plane on a cylinder of radius $r = E^{1/2} R_0$. The case associated with *Euglena*'s shearing mechanism is

$$\Phi(u, v) = R_{e_3}^{\phi} (ue_1 + (v + \gamma u)e_2), \quad g = \begin{bmatrix} 1 + \gamma^2 & \gamma \\ \gamma & 1 \end{bmatrix}, \tag{1.45}$$

where $R_{e_3}^{\phi}$ is a rotation with axis e_3 and angle $\phi = \tan^{-1}(\gamma)$.

The case associated with with *Lacrymaria*'s stretching mechanism is given instead by

$$\Phi(u, v) = \lambda u e_1 + \mu v e_2, \quad g = \begin{bmatrix} \lambda^2 & 0 \\ 0 & \mu^2 \end{bmatrix}. \tag{1.46}$$

The way surfaces deform as a consequence of metrics of type (1.45) is studied extensively in [17, 18], to which the reader is referred. A similar analysis can be done for metrics of type (1.46). Substituting (1.46) into (1.40) we obtain

$$\begin{bmatrix} (r/R_0)^2 & r^2 \psi'/R_0 \\ r^2 \psi'/R_0 & r'^2 + z'^2 + r^2 \psi'^2 \end{bmatrix} = \begin{bmatrix} \lambda^2 & 0 \\ 0 & \mu^2 \end{bmatrix} = \text{constant}, \tag{1.47}$$

which gives $r = \lambda R_0$, while the functions $\psi(v)$ and $z(v)$ are determined by solving the differential equations

$$\psi' = 0, \tag{1.48}$$

and

$$z' = \pm \mu. \tag{1.49}$$

Hence, by setting the integration constants $\psi(0)$ and $z(0)$ equal to zero and selecting the plus sign in (1.49) (solutions to (1.47) are determined up to a rigid motion allowing for translations along e_3, rotations about e_3, and $\pm$ inversion along e_3, which are here fixed), we have $\psi(v) = 0$ and $z(v) = \mu v$.

Having in mind the arrangement of microtubules in *Lacrymaria*'s envelope, and also the arrangement of fibers in the braided sheaths of McKibben artificial muscles, we are interested in the conformational changes of networks of helical material curves on S_0, when a metric change $g_0 \mapsto g$ transforms S_0 into S. We consider

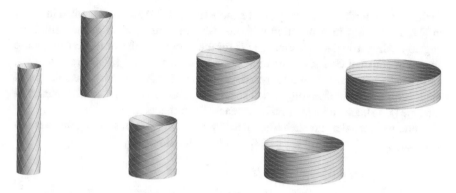

Fig. 1.7 Cylindrical surfaces obtained from a referential cylinder with $H_0/R_0 = 5$ by exploiting the area preserving stretching morphing principle (1.35) for $\lambda = \{0.75, 1, 1.5, 2, 2.5, 3\}$. Cylindrical surfaces are decorated by blue and yellow material fibers for $\theta_0 = \pi/4$ and $N = 10$. Figure reproduced from [36]

the $2N$ lines

$$u^{(k)}(v) = k\frac{2\pi}{N}R_0 \pm \tan(\vartheta_0)v\,, \quad k = 0, \ldots, N-1 \tag{1.50}$$

and their images in the reference and deformed configurations

$$\begin{aligned}\chi_0^{(k)}(v) &= \chi_0(u^{(k)}(v), v) \\ &= \left\{ R_0\cos\left(\frac{1}{R_0}u^{(k)}(v)\right), R_0\sin\left(\frac{1}{R_0}u^{(k)}(v)\right), v \right\}, \quad v \in (0, H_0)\end{aligned} \tag{1.51}$$

and

$$\begin{aligned}\chi^{(k)}(v) &= \chi(u^{(k)}(v), z(v)) \\ &= \left\{ \lambda R_0\cos\left(\frac{1}{R_0}u^{(k)}(v)\right), \lambda R_0\sin\left(\frac{1}{R_0}u^{(k)}(v)\right), \mu v \right\}, \quad v \in (0, H_0).\end{aligned} \tag{1.52}$$

Curves (1.51) are circular helices with radius R_0, screw axis parallel to e_3, and pitch angle ϑ_0. Curves (1.52) are circular helices with radius λR_0, screw axis parallel to e_3, and pitch angle $\tan^{-1}(\lambda\tan(\vartheta_0)/\mu)$ (when $\vartheta_0 = \pi/4$, the angular pitch of (1.52) is $\tan^{-1}(\lambda/\mu)$). These are all illustrated in Fig. 1.7.

1.6.4   Axisymmetric Surfaces with Non-constant Metric

We turn now to more general axisymmetric surfaces S of the form (1.39), obtained by axisymmetric actuation, i.e., by a non-constant metric g depending only on the "vertical" coordinate v (the coordinate along the symmetry axis) and not on the "azimuthal" coordinate u. The case of simple shear (1.33) has been studied in detail in [17, 18, 84, 85] in connection with the morphing mechanism of the pellicle of euglenids. In particular, we refer to [17] for the complete atlas of the axisymmetric shapes of constant Gaussian curvature surfaces (cylinders, cones, spheres, spindles, and pseudo-spheres) achievable by axisymmetric shearing, and for the solution of the inverse problem of finding which shear actuation patterns, i.e., which metric of the type (1.45), are capable of realizing each given shape.

Here, we consider the stretching metric given by (1.35), restricting attention to the area-preserving case of $\lambda\mu = 1$ (pure shear) for simplicity. From

$$\begin{bmatrix} (r/R_0)^2 & r^2\psi'/R_0 \\ r^2\psi'/R_0 & r'^2 + z'^2 + r^2\psi'^2 \end{bmatrix} = \begin{bmatrix} \lambda^2 & 0 \\ 0 & 1/\lambda^2 \end{bmatrix}, \tag{1.53}$$

where $\lambda = \lambda(v)$ and a prime denotes derivative with respect to v, we obtain

$$r(v) = \lambda(v)R_0, \tag{1.54}$$

$$\psi'(v) = 0, \tag{1.55}$$

and

$$z'(v) = \pm\frac{1}{\lambda(v)}\sqrt{1 - (R_0\lambda\lambda')^2}, \tag{1.56}$$

which can be solved with real $z(v)$ provided that

$$\lambda|\lambda'| \leq \frac{1}{R_0}. \tag{1.57}$$

This is a necessary condition for a metric of the form (1.35) to be realized by an axisymmetric surface (embeddability in $\mathbb{R}^3$).

We start by seeking surfaces of zero Gaussian curvature $K = 0$, more general than the cylinders of the previous section, which arise when $\lambda' = 0$. We thus have

$$0 = -(\lambda\lambda')' = -\frac{1}{2}(\lambda^2)'', \tag{1.58}$$

which implies that $\lambda(v)\lambda'(v) = C$, a constant such that $|C| \leq 1/R_0$. Hence,

$$z'(v) = \frac{dz}{dv} = \pm\frac{1}{\lambda(v)}\sqrt{1 - R_0^2C^2} \tag{1.59}$$

and, using (1.54), we deduce that

$$\frac{dr}{dz} = \frac{\frac{dr}{dv}}{\frac{dz}{dv}} = \pm\frac{R_0\lambda(v)\lambda'(v)}{\sqrt{1 - R_0^2C^2}} = \pm\sqrt{\frac{R_0^2C^2}{1 - R_0^2C^2}} =: \pm\tan(\phi), \tag{1.60}$$

which shows that S is a cone with axis parallel to e_3 and opening angle ϕ (measured clockwise from the e_3 axis). This angle tends to zero when $C \to 0$, and to $\pi/2$ when $C \to \pm 1/R_0$.

Moreover, it follows from (1.58) that $\lambda^2(v)$ is a linear function of $v \in (0, H_0)$. Thus, we can write λ as

$$\lambda = \tilde{\lambda}(\xi) = \sqrt{A(1-\xi) + B\xi}, \quad \xi := \frac{v}{H_0} \in (0,1), \quad \tilde{\lambda}(0) = \sqrt{A}, \quad \tilde{\lambda}(1) = \sqrt{B}$$

$$(1.61)$$

and integrate (1.59) to obtain

$$z(v) = z(0) \pm H_0 \sqrt{1 - R_0^2 C^2} \int_0^{v/H_0} \frac{dx}{\sqrt{A(1-x) + Bx}}. \tag{1.62}$$

Furthermore, from $\lambda\lambda' = (\lambda^2)'/2 = C$, we obtain

$$C = \frac{B - A}{2H_0}, \tag{1.63}$$

so that the embeddability condition $|C| \le 1/R_0$ is equivalent to

$$|B - A| - \frac{2H_0}{R_0} \le 0. \tag{1.64}$$

Setting all the integration constants to zero and choosing the positive sign in the previous formulas, we obtain the parametrization of S as

$$\lambda(v) = \sqrt{A\left(1 - \frac{v}{H_0}\right) + B\frac{v}{H_0}}, \tag{1.65}$$

$$r(v) = \lambda(v)R_0, \tag{1.66}$$

$$\psi(v) = 0, \tag{1.67}$$

and

$$z(v) = \frac{\sqrt{A(1 - R_0^2 C^2)}}{C}\left(\sqrt{1 - \frac{v}{H_0} + \frac{B}{A}\frac{v}{H_0}} - 1\right), \tag{1.68}$$

which describes a truncated cone with opening angle given by (1.60) and radii at the rim of the surface equal to $r(0) = R_0\sqrt{A}$ and $r(H_0) = R_0\sqrt{B}$. Here $A = \lambda^2(0)$, $B = \lambda^2(H_0)$, $C = (B - A)/(2H_0)$ and A, B must be such that $|B - A| - 2H_0/R_0 \le 0$.

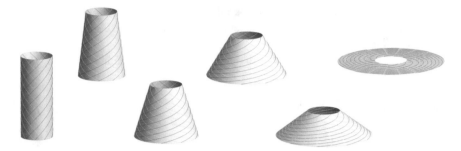

Fig. 1.8 Truncated cones obtained from a referential cylinder with $H_0/R_0 = 5$ by exploiting the stretching morphing principle (1.35) for $r(0)/R_0 = \{1, 1.5, 2, 2.5, 3, 3.32\}$ and $r(H_0)/R_0 = 1$. Conical surfaces are decorated by blue and yellow material fibers for $\theta_0 = \pi/4$ and $N = 10$. Figure reproduced from [36]

The images of the circular helices (1.51) after deformation are obtained by substituting the previous formulas into

$$\chi^{(k)}(v) = \chi(u^{(k)}(v), z(v))$$

$$= \left\{ r(v) \cos\left(\frac{1}{R_0} u^{(k)}(v)\right), r(v) \sin\left(\frac{1}{R_0} u^{(k)}(v)\right), z(v) \right\}, \quad v \in (0, H_0)$$

$$(1.69)$$

and are represented in Fig. 1.8.

Figure 1.9 shows a comparison between surfaces and networks of material lines arising form the two different actuation mechanisms. The figure shows that the same shapes can be obtained with two different metrics (one corresponding to a stretching mechanism, the other one to a shearing mechanism). The two mechanisms (not the shapes) are distinguishable because they lead to different deformations of the networks of material lines, and lead to different displacements of material points on the surfaces. On the left part of the figure, red lines suggest the evolution of the orientation of the pellicle strips of Euglenids, which deform according to the

Fig. 1.9 A comparison between identical shapes (cylinders and truncated cones) obtained by means of either a shearing (left) or stretching (right) mechanism. All the shapes are for $H_0/R_0 = 5$. The shorted cylinders correspond to $\gamma = \sqrt{3}$ (shearing mechanism) and to $\lambda = 2$ (stretching mechanism). Cones are both such that $r(0)/R_0 = 2.5$ and $r(H_0)/R_0 = 1$. Surfaces are decorated by colored material fibers to highlight the *embodiment* of the morphing principle and to emphasize the difference between the two morphing mechanisms. Figure reproduced from [36]

shearing mechanism (1.33). On the right side of the figure the blue and yellow lines suggest the evolution of the orientation of the network of threads in the sheath of a McKibben artificial muscle or of the microtubules in the cell envelope of *Lacrymaria*, which deform according to the stretching mechanism (1.35).

We remark that the theory of shape-shifting surfaces va Gaussian morphing described in this section is purely geometric. It is clear, however, that the shape-programming path in shape space we have described may require elastic deformations of the structural elements (rod, membrane, plate, shell, and block elements) making up the morphing surface. Since these elements will be stretched and bent, the way a morphing principle works when implemented into a concrete organism or in a concrete engineered device cannot be predicted without the explicit appreciation of its *embodiment* into a specific body architecture (in the case of organisms) or into specific arrangement of structural elements in the case of engineered devices. In other words, the way a morphing principle works in practice is crucially affected by the mechanical compliance of the materials and mechanisms that build the body or the device.

Thus, the difference between the two mechanisms illustrated in Fig. 1.9 is not without subtleties. For each fixed value γ of the shearing metric g, there is a stretching metric delivering the same shape, through stretches along different coordinate lines, the ones along the γ-dependent eigenvectors of g. A change of coordinates transforms one metric into the other. If, however, we insist on the fact that some curves in the reference configuration have the character of material lines, and that the embodiment of the shape-shifting mechanism governs the change of lengths and angles of material lines, then the two mechanisms by shearing and stretching are no longer interchangeable. Indeed, material lines are deformed in different ways by the two mechanisms and the embodiment reveals the difference between the two. Along the one-parameter family of shapes illustrated in Fig. 1.9, material lines initially straight and vertical deform at constant length by differential lateral sliding in the shearing mechanism (shown in the left). Instead, they remain straight and are shortened by the stretching mechanism (shown in the right). A more complete discussion of this issue, and a more complete characterization of shapes that can be produced with the two mechanisms (direct problem), and of the actuation patterns needed to obtain them in specific embodiments of the underlying Gaussian morphing principle (inverse problem) will be provided elsewhere.

1.6.5   Protruding Necks and Localized Bulges

Of particular interest in biology and soft robotics is the construction of localized bulges and of protruding necks starting from an axisymmetric shape. These are commonly exhibited by biological organisms and one can think of exploiting them in artificial devices for robotic manipulation, drug delivery, surgical tools for minimally invasive surgery, etc. In the context of the shearing mechanism (1.41), they have been used to mimic the peristaltic waves and traveling bulges rationalizing metaboly in Euglenids, see [17, 85]. In the context of the stretching mechanism, (1.42) or (1.43), they can be used to mimic the neck protrusion that the unicellular predator *Lacrymaria olor* uses to explore the environment to search for preys and to feed on them.

Indeed, let's consider (1.43), with the axially-symmetric stretch λ given in terms of a non-dimensional coordinate ξ along the symmetry axis by

$$\tilde{\lambda}(\xi) = 1 + A \left(1 - \exp\left[-\left(\frac{\xi - \xi_0}{D} \right)^2 \right] \right), \quad \xi := \frac{v}{H_0} \in (-1, 1), \qquad (1.70)$$

where $A > -1$ determines the strength of the bulge perturbation with respect to the reference cylinder (1.38), $D > 0$ determines its non-dimensional breadth, while $\xi_0 \in (-1, 1)$ locates the centre of the bulge along the symmetry axis.

The shapes resulting from (1.70) are shown in Fig. 1.10. They are closely reminiscent of the patterns exhibited by *Lacrymaria* when it protrudes its neck for

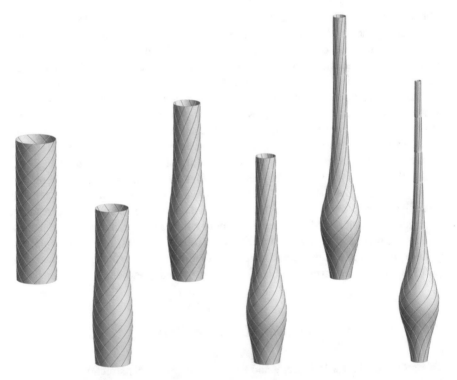

Fig. 1.10 Deformations resembling the neck extension of *Lacrymaria olor* obtained by means of the stretching mechanism (1.43) from a reference cylinder. Shapes are obtained using Eq. (1.70) with $A = \{0, -0.17, -0.34, -0.51, -0.68, -0.85\}$, $D = 0.28$, and $\xi_0 = 0.3$. The referential cylinder is such that $H_0/R_0 = 7$. Surfaces are decorated by blue and yellow material fibers defined by Eq. (1.50) with $\theta_0 = \pi/4$ and $N = 10$. Figure reproduced from [36]

feeding, see [39]. Shape changes of a similar kind are also exhibited by *Euglena* when it executes metaboly, even though the extent of the protrusions is less extreme. There, the shape-shifting mechanism (the metric change) and the displacements of material points on the surface (the sliding of the pellicle strips) arise from a continuous one–parameter family of shears (1.41) rather than stretches (1.43), see [17, 85]. The molecular and bio-physical details by which the two unicellular organisms control their behavior are still largely unknown.

1.7 Discussion and Outlook

In these notes we have followed two main threads, which are obviously intertwined.

On the one hand, we have considered biological and bio-inspired locomotion at small length scales, and swimming motility of microscopic unicellular organisms in particular. How does an organism or an artificial device move as a consequence of its mechanical interactions with the environment, if is shape varies in time? E.g., according to a periodic time-history? The main tool we have used in our mathematical approach to this navigation problem is geometric control theory. In this way, reaching position B from position A becomes a question of controllability, and doing so optimally (e.g., at minimal energetic cost) is a problem of optimal control.

On the other hand, we have considered the problem of how to control the shape of the outer envelope of biological organism or of a thin engineering structure. I.e., how can we morph surfaces according to a specified shape programming scheme? In particular, we have discussed the Gaussian morphing scheme, by which shape changes result form changes of Gaussian curvature, which are in turn induced by (active) lateral modulation of the metric of the surface.

There are cases, such as the metaboly of Euglenids, where the two threads are intertwined. Indeed, Euglena gracilis cells control the shape of their outer envelope by patterns of active pellicle shears, which represent modulations of the surface metric by surface shears.

This brings us to the issue of the practical use of the algorithmic morphing principle we have discussed, be it Gaussian morphing or more general morphing schemes, for concrete applications. Invariably, workers in the field of micro-swimmers like to mention the goal of using these principles to fabricate microscopic medical devices (micro-robots) for autonomous navigation inside the human body, capable of executing diagnostic, therapeutic, drug delivery, and minimally invasive surgery functions, in a concrete application of the visionary sequences of the movie 'Fantastic Voyage', or of the visionary words of R. Feynman, who was talking about surgeons that can be swallowed. More generally, robotics has shown concrete interest in concepts similar to the ones described in these notes, opening up the new field of Soft Robotics [66]. In this new paradigm, robots that are asked to interact with unstructured and rapidly changing environments can exploit the adaptability coming from compliant bodies, instead of relying only on complex control algorithms for more traditional architectures based on rigid body parts.

The community of Soft Robotics is expanding, but a key challenge to the actual deployment of robotic devices made at least in part of soft component comes from the availability of soft active materials with the right properties. Hydrogels have been used in several applications, in which actuation by solvent uptake is a feasible mechanism (see e.g. [5, 65, 67, 91] and the references cited therein). Liquid crystal elastomers (LCEs) [104] have also been identified as a very promising material, thanks to the availability of many stimuli to which they can respond to (temperature, light, electromagnetic fields, solvent uptake), and recent advances in manufacturing (that allow for a very fine resolution of the patterns of programmed director fields, which can be used to induce shape changes following Gaussian morphing principle). There is a large literature on modeling the mechanical response of LCE sheets, see [4, 7, 19, 26, 37, 43, 45, 46, 52, 78, 82, 93, 104–106] and the references cited therein for a small sample.

These materials pose serious challenges also at the level of mathematical modeling. Modeling the mechanics of phase transforming polymeric materials such as LCEs is highly nontrivial, since they exhibit nonlinearities in their response, for example due to domain evolution and hysteresis. Modeling these phenomena still represent a challenge, in spite of the fact that these topics having been in the forefront of research for many years, see e.g., [23, 42, 43]. Recent spectacular results on the possibility of producing low hysteresis shape memory materials, based on mathematical models (with low hysteresis behavior occurring when special compatibility relations between the metric changes in two neighboring domains are satisfied) testifies, however, that progress in this difficult subject is being made [57].

Acknowledgments We gratefully acknowledge the support by the European Research Council through the ERC Advanced Grant 340685-MicroMotility. These lecture notes draw freely from results obtained with several co-authors over the last 10 years, and published in the papers referenced in the bibliography (in particular, references [35, 36]). The collaboration with them has been a source of endless joy and inspiration.

References

1. D. Agostinelli, F. Alouges, A. De Simone, Peristaltic waves as optimal gaits in metameric bio-inspired robots. Front. Robot. and AI **5**, 99 (2018)
2. D. Agostinelli, A. Lucantonio, G. Noselli, A. DeSimone, Nutations in growing plant shoots: the role of elastic deformations due to gravity loading. J. Mech. Phys. Solids **136**, 103702 (2019)
3. D. Agostinelli, R. Cerbino, J. Del Alamo, A. DeSimone, A. Hoehn, C. Micheletti, G. Noselli, E. Sharon, J. Yeomans, Micromotility: state of the art, recent accomplishments and perspectives on the mathematical modeling of bio-motility at microscopic scales. Math. Eng. (2020) https://doi.org/10.3934/mine.2020011
4. V. Agostiniani, A. DeSimone, Γ-convergence of energies for nematic elastomers in the small strain limit. Contin. Mech. Thermodyn. **23**(3), 257–274 (2011)

5. V. Agostiniani, A. DeSimone, A. Lucantonio, D. Lucic, Foldable structures made of hydrogel bilayers. Math. Eng. **1**, 204–223 (2018). https://doi.org/10.3934/mine.2018.1.204
6. H. Aharoni, Y. Abraham, R. Elbaum, E. Sharon, R. Kupferman, Emergence of spontaneous twist and curvature in non-Euclidean rods: application to *Erodium* plant cells. Phys. Rev. Lett. **108**, 238106 (2012)
7. H. Aharoni, E. Sharon, R. Kupferman, Geometry of thin nematic elastomer sheets. Phys. Rev. Lett. **113**, 257801 (2014)
8. B. Alberts, A. Johnson, J. Lewis, D. Morgan, M. Raff, K. Roberts, P. Walter, *Molecular Biology of the Cell*, 6th ed. (Garland Science, New York, 2014)
9. F. Alouges, A. DeSimone, A. Lefebvre, Optimal strokes for low Reynolds number swimmers: an example. J. Nonlinear Sci. **18**(3), 277–302 (2008)
10. F. Alouges, A. DeSimone, A. Lefebvre, Optimal strokes for axisymmetric microswimmers. Eur. Phys. J. E **28**(3), 279–284 (2009)
11. F. Alouges, A. Desimone, L. Heltai, Numerical strategies for stroke optimization of axisymmetric microswimmers. Math. Models Methods Appl. Sci. **21**(2), 361–387 (2011)
12. F. Alouges, A. DeSimone, L. Giraldi, M. Zoppello, Self-propulsion of slender microswimmers by curvature control: N-link swimmers. Int. J. Non Linear Mech. **56**, 132–141 (2013)
13. F. Alouges, A. DeSimone, L. Heltai, A. Lefebvre, B. Merlet, Optimally swimming stokesian robots. Discrete Contin. Dynam. Systems B **18**, 1189–1215 (2013)
14. F. Alouges, A. DeSimone, L. Giraldi, M. Zoppello, Can magnetic multilayers propel artificial microswimmers mimicking sperm cells? Soft Rob. **2**, 117–128 (2015)
15. F. Alouges, A. Desimone, L. Giraldi, Y. Or, O. Wiezel, Energy-optimal strokes for multi-link microswimmers: Purcell's loops and Taylor's waves reconciled. New J. Phys. **21**(4), 043050 (2019)
16. D. Ambrosi, M.B. Amar, C.J. Cyron, A. DeSimone, A. Goriely, J.D. Humphrey, E. Kuhl, Growth and remodelling of living tissues: perspectives, challenges and opportunities. J. R. Soc. Interface **16**(157), 20190233 (2019)
17. M. Arroyo, A. DeSimone, Shape control of active surfaces inspired by the movement of euglenids. J. Mech. Physics Solids **62**, 99–112 (2014)
18. M. Arroyo, D. Milan, L. Heltai, A. DeSimone, Reverse engineering the euglenoid movement. Proc. Nat. Acad. Sci. USA **109**, 17874–17879 (2012)
19. M. Barchiesi, A. DeSimone, Frank energy for nematic elastomers: a nonlinear model. ESAIM Control Optim. Calc. Var. **21**(2), 372–377 (2015)
20. H.C. Berg, D.A. Brown, Chemotaxis in escherichia coli analysed by three-dimensional tracking. Nature **239**(5374), 500–504 (1972)
21. M. Bergert, A. Erzberger, R.A. Desai, I.M. Aspalter, A.C. Oates, G. Charras, G. Salbreux, E.K. Paluch, Force transmission during adhesion-independent migration. Nat. Cell Biol. **17**(4), 524 (2015)
22. L. Berti, L. Giraldi, C. Prud'Homme, Swimming at Low Reynolds Number. ESAIM: Proceedings and Surveys, EDP Sciences, 2019, pp. 1–10
23. P. Bladon, E.M. Terentjev, M. Warner, Transitions and instabilities in liquid crystal elastomers. Phys. Rev. E **47**, R3838–R3840 (1993)
24. A. Bressan, Impulsive control of lagrangian systems and locomotion in fluids. Discrete Contin. Dynam. Systems **20**(1), 1 (2008)
25. A.B.C. Buskermolen, H. Suresh, S.S. Shishvan, A. Vigliotti, A. DeSimone, N.A. Kurniawan, C.V.C. Bouten, V.S. Deshpande, Entropic forces drive cellular contact guidance. Biophys. J. **116**(10), 1994–2008 (2019)
26. M. Camacho-Lopez, H. Finkelmann, P. Palffy-Muhoray, M. Shelley, Fast liquid-crystal elastomer swims into the dark. Nat. Mater. **3**(5), 307 (2004)
27. L. Cardamone, A. Laio, V. Torre, R. Shahapure, A. DeSimone, Cytoskeletal actin networks in motile cells are critically self-organized systems synchronized by mechanical interactions. Proc. Natl. Acad. Sci. **108**(34), 13978–13983 (2011)

28. D. Carroll, E. Hankins, E. Kose, I. Sterling, A survey of the differential geometry of discrete curves. Math. Intell. **36**(4), 28–35 (2014)
29. A.N. Caruso, A. Cvetkovic, A. Lucantonio, G. Noselli, A. DeSimone, Spontaneous morphing of equibiaxially pre-stretched elastic bilayers: the role of sample geometry. Int. J. Mech. Sci. **149**, 481–486 (2018)
30. B. Chan, N.J. Balmforth, A.E. Hosoi, Building a better snail: lubrication and adhesive locomotion. Phys. Fluids **17**(11), 113101 (2005)
31. G. Charras, E. Paluch, Blebs lead the way: how to migrate without lamellipodia. Nat. Rev. Mol. Cell Biol. **9**(9), 730 (2008)
32. S. Childress, *Mechanics of Swimming and Flying*, vol. 2 (Cambridge University Press, Cambridge, 1981)
33. G. Cicconofri, A. DeSimone, A study of snake-like locomotion through the analysis of a flexible robot model. Proc. R. Soc. London, Ser. A Math. Phys. Eng. Sci. **471**(2184), 20150054 (2015)
34. G. Cicconofri, A. DeSimone, Motion planning and motility maps for flagellar microswimmers. Eur. Phys. J. E **39**, 72 (2016)
35. G. Cicconofri, A. DeSimone, Modelling biological and bio-inspired swimming at microscopic scales: recent results and perspectives. Comput. Fluids **179**, 799–805 (2019)
36. G. Cicconofri, M. Arroyo, G. Noselli, A. DeSimone, Morphable structures from unicellular organisms with active, shape-shifting envelopes: variations on a theme by Gauss. Int. J. Non Linear Mech. **118**, 103278 (2020)
37. S. Conti, A. DeSimone, G. Dolzmann, Soft elastic response of stretched sheets of nematic elastomers: a numerical study. J. Mech. Phys. Solids **50**(7), 1431–1451 (2002)
38. G. Corsi, A. DeSimone, C. Maurini, S. Vidoli, A neutrally-stable shell in a Stokes flow: a rotational Taylor sheet. Proc. R. Soc. A **475**, 20190178 (2019)
39. S.M. Coyle, E.M. Flaum, H. Li, D. Krishnamurthy, M. Prakash, Coupled active systems encode emergent behavioral dynamics of the unicellular predator Lacrymaria olor. Curr. Biol. **29**(22), 3838–3850. e3 (2019)
40. G. Dal Maso, A. DeSimone, M. Morandotti, An existence and uniqueness result for the motion of self-propelled micro-swimmers. SIAM J. Math. Anal. **43**, 1345–1368 (2011)
41. C. Darwin, *The Power of Movement in Plants* (John Murray, London, 1880)
42. A. DeSimone, Hysteresis and imperfection sensitivity in small ferromagnetic particles. Meccanica **30**(5), 591–603 (1995)
43. A. DeSimone, Energetics of fine domain structures. Ferroelectrics **222**(1–4), 275–284 (1999)
44. A. DeSimone, Spontaneous bending of pre-stretched bilayers. Meccanica **53**, 511–518 (2018)
45. A. DeSimone, G. Dolzmann, Macroscopic response of nematic elastomers via relaxation of a class of $SO(3)$-invariant energies. Arch. Ration. Mech. Anal. **161**(3), 181–204 (2002)
46. A. DeSimone, L. Teresi, Elastic energies for nematic elastomers. Eur. Phys. J. E **29**(2), 191–204 (2009)
47. A. DeSimone, A. Tatone, Crawling motility through the analysis of model locomotors: two case studies. Eur. Phys. J. E **35**(9), 85 (2012)
48. A. DeSimone, F. Guarnieri, G. Noselli, A. Tatone, Crawlers in viscous environments: linear vs non-linear rheology. Int. J. Non Linear Mech. **56**, 142–147 (2013)
49. A. DeSimone, P. Gidoni, G. Noselli, Liquid crystal elastomer strips as soft crawlers. J. Mech. Phys. Solids **84**, 254–272 (2015)
50. M.P. do Carmo, *Differential Geom. of Curves and Surfaces*. Paperback (Prentice-Hall, Inc, Englewood Cliffs, 1976)
51. K. Drescher, R.E. Goldstein, N. Michel, M. Polin, I. Duval, Direct measurement of the flow field around swimming microorganisms. Phys. Rev. Lett. **105**, 168101 (2010)
52. A. Fukunaga, K. Urayama, T. Takigawa, A. DeSimone, L. Teresi, Dynamics of electro-opto-mechanical effects in swollen nematic elastomers. Macromolecules **41**(23), 9389–9396 (2008)
53. E.A. Gaffney, H. Gadelha, D.J. Smith, J.R. Blake, J.C. Kirkman-Brown, Mammalian sperm motility: observation and theory. Annu. Rev. Fluid Mech. **43**, 501–528 (2011)

54. P. Gidoni, G. Noselli, A. DeSimone, Crawling on directional surfaces. Int. J. Non Linear Mech. **61**, 65–73 (2014)
55. N. Giuliani, N. Heltai, A. DeSimone, Predicting and optimizing micro-swimmer performance from the hydrodynamics of its components: the relevance of interactions. Soft Rob. **5**(4) (2018)
56. R.E. Goldstein, Green algae as model organisms for biological fluid dynamics. Annu. Rev. Fluid Mech. **47**, 343–375 (2015)
57. H. Gu, L. Bumke, C. Chluba, E. Quandt, R.D. James, Phase engineering and supercompatibility of shape memory alloys. Mater. Today **21**(3), 265–277 (2018)
58. J.S. Guasto, K.A. Johnson, J.P. Gollub, Oscillatory flows induced by microorganisms swimming in two dimensions. Phys. Rev. Lett. **105**, 168102 (2010)
59. Z.V. Guo, L. Mahadevan, Limbless undulatory propulsion on land. Proc. Natl. Acad. Sci. **105**(9), 3179–3184 (2008)
60. P. Holmes, R.J. Full, D. Koditschek, J. Guckenheimer, The dynamics of legged locomotion: models, analyses, and challenges. SIAM Rev. **48**(2), 207–304 (2006)
61. D.L. Hu, J. Nirody, T. Scott, M.J. Shelley, The mechanics of slithering locomotion. Proc. Natl. Acad. Sci. **106**(25), 10081–10085 (2009)
62. A.J. Ijspeert, Central pattern generators for locomotion control in animals and robots: a review. Neural Netw. **21**(4), 642–653 (2008)
63. A.Y. Khapalov, Local controllability for a swimming model. SIAM J. Control. Optim. **46**(2), 655–682 (2007)
64. A. Khapalov, P. Cannarsa, F.S. Priuli, G. Floridia, Well-posedness of 2-d and 3-d swimming models in incompressible fluids governed by navier–stokes equations. J. Math. Anal. Appl. **429**(2), 1059–1085 (2015)
65. J. Kim, J.A. Hanna, M. Byun, C.D. Santangelo, R.C. Hayward, Designing responsive buckled surfaces by halftone gel lithography. Science **335**(6073), 1201–1205 (2012)
66. S. Kim, C. Laschi, B. Trimmer, Soft robotics: a bioinspired evolution in robotics. Trends Biotechnol. **31**(5), 287–294 (2013)
67. Y. Klein, E. Efrati, E. Sharon, Shaping of elastic sheets by prescription of non-Euclidean metrics. Science **315**, 1116–1120 (2007)
68. J.H. Lai, J.C. del Alamo, J. Rodríguez-Rodríguez, J.C. Lasheras, The mechanics of the adhesive locomotion of terrestrial gastropods. J. Exp. Biol. **213**(22), 3920–3933 (2010)
69. E. Lauga, A.E. Hosoi, Tuning gastropod locomotion: modeling the influence of mucus rheology on the cost of crawling. Phys. Fluid. **18**(11), 113102 (2006)
70. E. Lauga, T.R. Powers, The hydrodynamics of swimming microorganisms. Rep. Prog. Phys. **72**(9), 096601 (2009)
71. B.S Leander, G. Lax, A. Karnkowska, A.G.B. Simpson, Euglenida. In: *Handbook of the Protists* (Springer, Cham, 2017)
72. Sir J. Lighthill. *Mathematical Biofluiddynamics* (SIAM, Philadelphia, 1975)
73. J. Lin, D. Nicastro, Asymmetric distribution and spatial switching of dynein activity generates ciliary motility. Science **360**(6387), eaar1968 (2018)
74. C.B. Lindemann, K.A. Lesich, Flagellar and ciliary beating: the proven and the possible. J. Cell Sci. **123**(4), 519–528 (2010)
75. J. Lohéac, J.-F. Scheid, M. Tucsnak, Controllability and time optimal control for low Reynolds numbers swimmers. Acta Appl. Math. **123**(1), 175–200 (2013)
76. K.E. Machin, Wave propagation along flagella. J. Exp. Biol. **35**(4), 796–806 (1958)
77. R. Marchello, M. Morandotti, H. Shum, M. Zoppello, The *n*-link swimmer in three dimensions: controllability and optimality results. arXiv preprint arXiv:1912.04998 (2019)
78. C.D. Modes, M. Warner, Negative Gaussian curvature from induced metric changes. Phys. Rev. E **92**, 010401 (2015)
79. F. Montenegro-Johnson, E. Lauga, Optimal swimming of a sheet. Phys. Rev. E **89**, 060701(R) (2014)
80. R. Montgomery, *A Tour of Subriemannian Geometries, Their Geodesics and Applications*, vol. 91 (American Mathematical Society, Providence, 2002)

81. A. Montino, A. DeSimone, Three-sphere low-Reynolds-number swimmer with a passive elastic arm. Eur. Phys. J. E **38**(42), 1–10 (2015)
82. C. Mostajeran, Curvature generation in nematic surfaces. Phys. Rev. E **91**, 062405 (2015)
83. A. Najafi, R. Golestanian, Simple swimmer at low Reynolds number: three linked spheres. Phys. Rev. E **69**(6), 062901 (2004)
84. G. Noselli, M. Arroyo, A. DeSimone, Smart helical structures inspired by the pellicle of euglenids. J. Mech. Phys. Solids **123**, 234–246 (2019)
85. G. Noselli, A. Beran, M. Arroyo, A. DeSimone, Swimming Euglena respond to confinement with a behavioral change enabling effective crawling. Nat. Phys. **15**, 496–502 (2019)
86. C. Pehlevan, P. Paoletti, L. Mahadevan, Integrative neuromechanics of crawling in D. melanogaster larvae. Elife **5**, e11031 (2016)
87. E.M. Purcell, Life at low Reynolds number. Am. J. Phys. **45**(1), 3–11 (1977)
88. P. Recho, T. Putelat, L. Truskinovsky, Contraction-driven cell motility. Phys. Rev. Lett. **111**(10), 108102 (2013)
89. P. Recho, A. Jerusalem, A. Goriely, Growth, collapse, and stalling in a mechanical model for neurite motility. Phys. Rev. E **93**(3), 032410 (2016)
90. M. Rossi, G. Cicconofri, A. Beran, G. Noselli, A. DeSimone, Kinematics of flagellar swimming in Euglena gracilis: Helical trajectories and flagellar shapes. Proc. Natl. Acad. Sci. U. S. A. **114**(50), 13085–13090 (2017)
91. C. Santangelo, Buckling thin disks and ribbons with non-Euclidean metrics. Europhys. Lett. **86**, 34003 (2011)
92. P. Sartori, V.F. Geyer, A. Scholich, F. Jülicher, J. Howard, Dynamic curvature regulation accounts for the symmetric and asymmetric beats of *Chlamydomonas* flagella. Elife **5**, e13258 (2016)
93. Y. Sawa, K. Urayama, T. Takigawa, A. DeSimone, L. Teresi, Thermally driven giant bending of liquid crystal elastomer films with hybrid alignment. Macromolecules **43**, 4362–4369 (2010)
94. P. Sens, Rigidity sensing by stochastic sliding friction. Europhys. Lett. **104**(3), 38003 (2013)
95. A. Shahaf, E. Efrati, R. Kupferman, E. Sharon, Geometry and mechanics in the opening of chiral seed pods. Science **333**(6050), 1726–1730 (2011)
96. D. Tam, A.E. Hosoi, Optimal stroke patterns for Purcell's three-link swimmer. Phys. Rev. Lett. **98**, 068105 (2007)
97. D. Tam, A.E. Hosoi, Optimal kinematics and morphologies for spermatozoa. Phys. Rev. E **83**, 045303(R) (2011)
98. G.I. Taylor, Analysis of the swimming of microscopic organisms. Proc. R. Soc. Lond. A Math. Phys. Eng. Sci. **209**, 447–461 (1951)
99. G.I. Taylor, Low Reynolds number flows, movie for The National Committee for Fluid Mechanics Films (1966). http://web.mit.edu/hml/ncfmf.html. Online accessed 6 Nov 2019
100. S. Timoshenko, Analysis of bi-metal thermostats. J. Optical Soc. Am. **11**, 233–255 (1925)
101. B. Tondu, Modelling of the McKibben artificial muscle: a review. J. Intell. Mater. Syst. Struct. **23**, 225–253 (2012)
102. Q. Wang, H. Othmer, The performance of discrete models of low Reynolds number swimmers. Math. Biosci. Eng. **12**, 1303 (2015)
103. Q. Wang, H.G. Othmer, Computational analysis of amoeboid swimming at low Reynolds number. J. Math. Biol. **72**(7), 1893–1926 (2016)
104. M. Warner, E.M. Terentjev, *Liquid Crystal Elastomers*, vol. 120 (Oxford University Press, Oxford, 2007)
105. M. Warner, C.D. Modes, D. Corbett, Curvature in nematic elastica responding to light and heat. Proc. R. Soc. Lond. A Math. Phys. Eng. Sci. **466**(2122), 2975–2989 (2010)
106. T.J. White, D.J. Broer, Programmable and adaptive mechanics with liquid crystal polymer networks and elastomers. Nat. Mat. **14**(11), 1087 (2015)

107. H. Wu, A. Farutin, W.-F. Hu, M. Thiébaud, S. Rafaï, P. Peyla, M.-C. Lai, C. Misbah, Amoeboid swimming in a channel. Soft Matt. **12**(36), 7470–7484 (2016)
108. S. Zhang, R.D. Guy, J.C. Lasheras, J.C. del Álamo, Self-organized mechano-chemical dynamics in amoeboid locomotion of physarum fragments. J. Phys. D. Appl. Phys. **50**(20), 204004 (2017)
109. J. Zhu, A. Mogilner, Mesoscopic model of actin-based propulsion. PLoS Comput. Biol. **8**(11), e1002764 (2012)

Chapter 2
Models of Cell Motion and Tissue Growth

Benoît Perthame

Abstract The mathematical description of cell movement, from the individual scale to the collective motion, is a rich and complex domain of biomathematics which leads to several types of questions and partial differential equations. For instance, bacteria move by run-and-tumble movement, which is well described, at the cell scale, by a kinetic equation in the phase coordinates. At the population scale, chemotactic effects lead to the famous parabolic Keller–Segel system, and the many improvements of it that have been addressed recently.

When considering living tissues, concepts issued from mechanics arise. Notions of pressure, phases, incompressibility are used in systems which carry the typical parabolic and hyperbolic characters of fluid mechanics. Their complexity is directly related to the details in the biological description and opens numerous mathematical questions which are poorly understood.

The various process involved in cell movements can be considered at the cell scale, at the population scale and, for tissues, at the organ scale. This leads to study singular perturbation problems of various types. For tumor growth, the tumor boundary can appear as a free boundary or as an internal layer.

2.1 Introduction

The mathematical description of cell movement from the individual scale to the collective motion, is a rich and complex domain of biomathematics which has been treated for a long time and with a rich variety of Partial Differential Equations. The goal, in this Chapter, is to give a flavor of the domain, showing the miscellaneous types of equations that can be encountered, the qualitative behavior of solutions and

B. Perthame (✉)
Sorbonne Université, CNRS, Université de Paris, Inria, Laboratoire Jacques-Louis Lions, Paris, France
e-mail: Benoit.Perthame@sorbonne-universite.fr

© The Editor(s) (if applicable) and The Author(s), under exclusive licence
to Springer Nature Switzerland AG 2020
A. DeSimone et al., *The Mathematics of Mechanobiology*,
Lecture Notes in Mathematics 2260, https://doi.org/10.1007/978-3-030-45197-4_2

some mathematical aspects. There are two general views which lead the line of the chapter:

- Present the domain of cell populations along the scales. We depart from the cell individual behavior, with the kinetic (mesoscopic) scale. This leads to the various forms of chemotaxis of swimming bacteria. Then, compressible models are presented as well as their incompressible limit. In some sense, this plan follows the traditional view of fluid mechanics.
- Explain how individual cells, which can be bacteria, interacting in a rather elementary way with their environment (secreting chemical substances and reacting to them) can generate collective behaviors which are complex and seem to result from organized strategies.

Several examples of reaction-diffusion equations, where the mechanism is due to interaction of cells with the environment through nutrients or other physical effects, are presented in [69] and in [50], showing mechanisms which underly the patten formation ability of such mechanisms. Reaction-diffusion equations are indeed usually encountered when considering populations of cells. However, many more types of equations, with very diverse nonlinearities can occur depending of the scale of interest.

At the individual scale, bacteria swim by a series of straight jumps followed by a fast reorganization of their flagella which lead to a new direction of jump. This trajectory is called *run-and-tumble* and is well described by a kinetic (linear Boltzman) equation in the phase space (position, velocity). A major issue here is to explain the modulation of these changes of direction depending on the environment. This environment may be characterized by the concentration of a chemoattractant and this results in a dynamics called chemotaxis where cells move preferentially toward the chemoattractant. Cells can emit the chemoattaracant themselves, and this leads to nonlinear systems.

Multiscale analysis allows to see the process at the population size and we arrive at a Fokker–Planck equation for cell number density coupled to the diffusion equation for the chemoattractant concentration. The resulting set of equations is called the Keller–Segel system. This famous system has attracted a lot of attention because of the complex patterns exhibited by solutions and in particular the blow-up phenomena which leads the cells to a pointwise concentration. Also, many improvements have been addressed recently which better fit some experimental observations.

When considering living tissues, which are denser ensembles of cells, fibers, liquids and molecules, these concepts are not enough, even if chemotaxis is important also. Then, the description uses concepts issued from mechanics and multiphase flows. Notions of pressure, Darcy's law, incompressibility are used in systems which carry parabolic and hyperbolic characters. Their complexity is directly related to the details in the biological description and opens numerous mathematical questions. A recent and comprehensive presentation of tissue mechanics can be found in [51]. The case of tumor growth has been attracted a huge literature recently and several surveys are available, [9, 46, 64, 85].

The domain of cell movement at the population scale is so vast that many aspects cannot be treated in this chapter. For instance, we do not touch the question of interaction between cell movements and the surrounding fluid, or chemotaxis in fluids [62] neither how cilia or flagella direct the motion of cells. The internal description of crawling cells are also wide subjects of present interest which are treated in the other chapters of this Lecture Note.

2.2 Bacterial Movement by Run and Tumble

Bacterial motion and bacterial population self-organisation is a wide and fascinating area of biology, which has generated an important literature with the progresses of experimental observations. A major question is to understand how simple cells can communicate and generate complex collective behaviors. The mathematical modeling should be able to reproduce the observations by numerical simulations.

The mechanism is due to the interaction of cells with their environment through nutrients or other physical effects [50, 69]. A specific example of communication mechanisms through the environment, used by bacteria, is chemotaxis, i.e., the motion of cells directed by a chemical signal. It is the central mechanism for *Escherichia coli*, which has raised an enormous interest since Adler's seminal paper [1], see also [18, 20, 67] and the book [10] for all biological aspects of *E. coli*.

Since the 80's, observations at the cell scale have shown that bacteria as *E. Coli* or *B. Subtilis* move by a series of straight jumps followed by a fast reorganization of their flagella which lead to a new direction of jumps, depending on the coordination of molecular motors that control the flagella. This is called the *run and tumble* movement. To give an idea of scales, the run time is about 1 s, the run length is a few μm and tumbling takes a much shorter time ($1/10 s$). To take into account that phenomena the kinetic formalism is needed and that was proposed, early after the first observations, by Alt and his co-authors [3, 71]. Recent surveys on the subject can be found in [26, 42, 54].

2.2.1 Modeling Run and Tumble

The modeling of run and tumble mechanism goes as follows. Denote by $f(x, \xi, t)$ the number density of cells located at $x \in \mathbb{R}^d$ and moving with the velocity $\xi \in V$, usually it is admitted that tumbles are always with the same speed and thus

$$V = \mathbb{S}^{d-1}.$$

From the knowledge of $f(x, \xi, t)$, it is usual in this domain to compute the macroscopic quantities as

$$n(x, t) := \int_V f(x, \xi, t) d\xi, \qquad\qquad \text{density,} \qquad (2.1)$$

$$n(x, t) u(x, t) := \int_V \xi f(x, \xi, t) d\xi, \qquad\qquad \text{momentum.} \qquad (2.2)$$

The physical representation of the run-and-tumble movement leads to write the equation

$$\begin{cases} \dfrac{\partial f(x, \xi, t)}{\partial t} + \overbrace{\xi \cdot \nabla_x f}^{\text{run}} = \overbrace{\mathcal{K}[c, f]}^{\text{tumble}}, \\[2mm] f(x, \xi, t = 0) = f^0(x, \xi) \geq 0, \qquad f^0 \in L^1. \end{cases} \qquad (2.3)$$

This equation is similar to scattering (see [7, 49] for instance) and the difficulty here is to take into account the rules leading cells to tumble, which are described by the term

$$\mathcal{K}[c, f] = \underbrace{\int_V K(c; \xi, \xi') f(x, \xi', t) d\xi'}_{\text{cells of velocity } \xi' \text{ turning to } \xi} - \underbrace{\int_V K(c; \xi', \xi) d\xi' \; f(x, \xi, t)}_{\text{cells of velocity } \xi \text{ turning to } \xi'} .$$

$$(2.4)$$

Here $c(x, t)$ describes a molecular environment which modulates cell responses. An important issue is that K may depend functionally on c. For instance it may depend in a non-local way on c, or on derivatives of c. To begin with, being given a function

$$\Phi \in C^1(\mathbb{R}; \mathbb{R}), \qquad 0 < \min \Phi(\cdot) < \max \Phi(\cdot) < \infty,$$

we can take a kernel with memory

$$K(c; \xi, \xi') = \Phi\big(c(x - \varepsilon\xi', t)\big), \qquad (2.5)$$

which expresses that a cell responds using the average concentration during their run of duration 2ε and using a middle rule for the integration. The function $\Phi(\cdot)$ takes into account possible response modulation to the signal. One may be more accurate and use an integration rule

$$K(c; \xi, \xi') = \Phi\big(\omega * c(x, t)\big)$$

for an appropriate kernel ω and convolution along the path.

This chemical signal $c(\cdot)$ can be emitted by the cells themselves and diffused in the media, then one writes

$$\tau \frac{\partial c}{\partial t} - \Delta c(x, t) = n(x, t) := \int_V f(x, \xi, t) d\xi, \tag{2.6}$$

with $\tau \geq 0$ the molecular diffusion time scale, usually small compared to cell dynamics. But the function $c(\cdot)$ can also be imposed from outside and then, Eq. (2.3) is linear.

The chemical signal $c(\cdot)$ can also be a nutrient consumed by the cells. Then we write, when the nutrient is not produced,

$$\tau \frac{\partial c}{\partial t} - \Delta c(x, t) + \lambda c(x, t) n(x, t) := 0, \tag{2.7}$$

with λ the consumption rate.

2.2.2   Existence of Solutions

According to the usual theory (see [49] for instance), some general properties of solutions of Eq. (2.3) are

1. *Non-negativity.* We have for all times $f(x, \xi, t) \geq 0$.
2. *Cell number conservation.* We have for all times $t \geq 0$

$$\int_{\mathbb{R}^d \times V} f(x, \xi, t) dx \, d\xi = \int_{\mathbb{R}^d \times V} f^0(x, \xi) dx \, d\xi.$$

Notice that this property follows from the symmetric form of the tumbling kernel where both $K(c; \xi, \xi')$ and $K(c; \xi', \xi)$ appear.

It is however difficult to draw more elaborate conclusions in terms of a priori bounds, in particular when the kinetic equation and the diffusion equation for chemical signal $c(c, t)$ are coupled. This difficulty opened the route to several existence results, which are still not complete in full generality and in particular when the tumbling kernel depends on ∇c, see [17, 33, 57]. Blow-up of solutions, under certain large mass conditions when the tumbling kernel depends on ∇c, has been obtained in [16].

Here we state a result from [33] which proves global existence of locally (in time) bounded solutions, thus extending a result in [53] in the linear case.

The existence theory for the nonlinear system (2.3)–(2.5) was settled in [33] and yields the

Theorem 2.1 ([33]) *In dimension $d = 3$, assume that V is bounded and that $f_0 \in L^\infty(\mathbb{R}^d \times V)$, then there is a unique solution to the system (2.3)–(2.5),*

$$f \in C\big([0, \infty); L^1(\mathbb{R}^d \times V)\big).$$

Moreover we have for all $T > 0$ and $0 \le t \le T$,

$$0 \le f(t, x, \xi) \le C(T),$$

$$\|\nabla c(t)\|_{L^p(\mathbb{R}^d)} \le C(T), \qquad \frac{d}{d-1} < p \le \infty,$$

$$\|c(t)\|_{L^p(\mathbb{R}^d)} \le C(T), \qquad d < p \le \infty.$$

for some constant $C(T)$,

The proof is based on dispersion estimates, [73].

This result provides global strong solutions and therefore shows a fundamental difference with the macroscopic model of Patlak/Keller–Segel system (see below) since the latter exhibits blow-up. This is rather counter-intuitive since we can expect that solutions to a hyperbolic equation, as the kinetic equation, has weaker estimates than the related parabolic Patlak/Keller–Segel equation.

The parabolic equation on c can be treated as well and several extensions of Theorem 2.1 have been obtained and for examples of specific dependency upon ∇c in the tumbling kernel K, see [16, 17, 57].

2.2.3 Derivation of the Patlak/Keller–Segel System

It is standard to derive macroscopic equations with only space variable from kinetic equations (in the phase space). To do so, one uses a rescaling in space and time to pass from the cell scale to the population scale. In particular the derivation allows a computation of the so-called "transport coefficients", i.e., the coefficients of the macroscopic model departing from the knowledge of the kinetic equation, i.e., the individual behavior. This has been a strong property, from the very beginning of kinetic theory, by Boltzmann, see [32] for instance. For linear scattering equations, the seminal paper [7] gives a complete proof including correctors. We explain the method for the run-and-tumble model at hand and the derivation of equations of Keller–Segel type [60].

Using the small memory parameter ε introduced in Eq. (2.5) for the tumbling rate, one can rescale space and time in the kinetic equation while keeping the diffusion equation for the chemoattractant (this is questionable and a more complete analysis should use the parameter range in the chemoattractant equation to justify the reasoning). Under the symmetry assumption on V that

$$\int_V \xi d\xi = 0, \tag{2.8}$$

we may use the diffusive rescaling (time scales as the square of space) and, following [53, 55, 72] for instance, we arrive at

$$\varepsilon \frac{\partial}{\partial t} f_\varepsilon(x, \xi, t) + \xi \cdot \nabla_x f_\varepsilon = \frac{1}{\varepsilon} \mathcal{K}[c_\varepsilon, f_\varepsilon], \tag{2.9}$$

$$f_\varepsilon(x, \xi, t = 0) = f^0(x, \xi) \geq 0, \qquad f^0 \in L^1. \tag{2.10}$$

As $\varepsilon \to 0$, and using the notation (2.1), the macroscopic limit is the non-linear Fokker–Planck equation (see again [33])

$$\frac{\partial}{\partial t} n(x, t) - \mathrm{div}[D(c)\nabla n] + \mathrm{div}[n\chi(c)\nabla c] = 0, \tag{2.11}$$

$$\tau \partial_t c(x, t) - \Delta c = n(x, t). \tag{2.12}$$

with the transport coefficients

$$D(c) = \frac{1}{|V|^2} \frac{\int_{V \times V} \xi \otimes \xi d\xi}{\Phi(c)}, \qquad \chi(c) = \frac{1}{|V|^2} \int_{V \times V} \xi \otimes \xi d\xi \frac{\Phi'(c)}{\Phi(c)}.$$

This system is called the Patlak/Keller–Segel system. It has been widely studied usually with D constant and χ depending on n, see Sect. 2.3 for variants and further results around this system.

Proof Let us show the formal derivation again, assuming convergence of all functions. First, we identify

$$\int_V [\Phi(c(x-\varepsilon\xi', t)) f_\varepsilon(x, \xi', t) - \Phi(c(x-\varepsilon\xi, t)) f_\varepsilon(x, \xi, t)] d\xi' = \varepsilon^2 \frac{\partial}{\partial t} f_\varepsilon(x, \xi, t) + \varepsilon\xi \cdot \nabla_x f_\varepsilon,$$

therefore as $\varepsilon \to 0$ we find

$$\int_V [\Phi(c(x, t)) f(x, \xi', t) - \Phi(c(x, t)) f(x, \xi, t)] d\xi' = 0$$

which means that the limit f of f_ε is independent of ξ and thus satisfies

$$f(x, \xi, t) = \frac{1}{|V|} n(x, t), \qquad \xi \in V.$$

As a second step we write, integrating in ξ,

$$\frac{\partial}{\partial t} n_\varepsilon(x, t) + \mathrm{div} J_\varepsilon(x, t) = 0, \qquad J_\varepsilon(x, t) = \frac{1}{\varepsilon} \int_V \xi f_\varepsilon(x, \xi, t) d\xi. \tag{2.13}$$

The third step is to compute J_ε which we do in using Eq. (2.10) as follows. After multiplication by a component ξ_j and integrating in ξ, we have

$$\frac{1}{\varepsilon} \int_{V \times V} [\Phi(c(x - \varepsilon \xi', t)) \xi_j f_\varepsilon(x, \xi', t) - \Phi(c(x - \varepsilon \xi, t)) \xi_j f_\varepsilon(x, \xi, t)] d\xi' d\xi$$

$$= \varepsilon \frac{\partial}{\partial t} \int_V \xi_j f_\varepsilon(x, \xi, t) d\xi + \int_V \xi \xi_j \cdot \nabla_x f_\varepsilon d\xi.$$

We neglect the terms in ε and get, using (2.8),

$$-\frac{1}{\varepsilon} \int_{V \times V} \Phi(c(x - \varepsilon \xi, t)) \xi_j f_\varepsilon(x, \xi, t)] d\xi' d\xi = \frac{1}{|V|} \int_V \xi \xi_j \cdot \nabla_x n d\xi + O(\varepsilon)$$

and a Taylor expansion of $\Phi(c(x - \varepsilon \xi, t))$ gives

$$-\Phi(c(x, t)) |V| J_\varepsilon(x, t) - |V| \Phi'(c) \nabla_x c(x, t) \int_{V \times V} \xi \xi_j d\xi = \frac{1}{|V|} \int_V \xi \xi_j \cdot \nabla_x n d\xi + O(\varepsilon).$$

This formula can be written as

$$J_\varepsilon(x, t) = -\frac{\Phi'(c)}{\Phi(c)} \nabla_x c(x, t) \int_{V \times V} \xi \xi_j d\xi - \frac{1}{|V|^2 \Phi(c)} \int_V \xi \xi_j d\xi \cdot \nabla_x n + O(\varepsilon).$$

Inserting this expression in (2.13) yields the result.

2.2.4 Modulation Along the Path

More realistic kinetic models of bacterial movement use the modulation of signal by *E. coli*. It turns out that bacteria increase the jump length when they feel an increasing chemotactic signal and reduce their jumps when the signal decreases along their path, [41, 65]. This leads to change the tumbling rule (2.5) to

$$K(c; \xi, \xi') = \Phi\left(\frac{\partial c}{\partial t} + \xi' . \nabla c\right). \tag{2.14}$$

Such models where proposed by Erban and Othmer [43], Dolak and Schmeiser [41], Erban and Othmer [44] and can be established from more precise descriptions of the movement which take into account internal molecular pathway, see [79, 89]. From these papers, we borrow the stiff response case, when

$$\Phi(z) = \begin{cases} k_- & \text{for } z < 0, \\ k_+ < k_- & \text{for } z > 0. \end{cases} \tag{2.15}$$

More generally $\Phi(\cdot)$ is a (smooth) decreasing function but stiffness is definitively a correct assumption

$$\Phi_\delta(z) = \Phi\left(\frac{z}{\delta}\right), \qquad \Phi(\pm\infty) = k_\pm, \tag{2.16}$$

for a 'small' constant $\delta > 0$.

An important aspect of taking into account the path derivative in (2.16) emerges from possible steady states and traveling pulse solutions which fit the experimental observations, [28, 86, 87].

The boundedness of the function Φ_δ makes that the difficulties for well-posedness are not the same here as for the tumbling rate of Sect. 2.2.2. They arise from the compactness of derivatives of c in the expression (2.14).We refer to the analysis in [75] for that aspect.

The macroscopic limit, using a tumbling kernel as (2.14), leads to a new class of macroscopic limits. called the *Flux Limited Keller-Segel system*. In [41], the hyperbolic limit is proved, leading to Eq. (2.22) in Sect. 2.3. In [75] another derivation is proved, with the diffusive scaling, using a different small parameter, namely the stiffness in response denoted by δ in (2.16).

2.3   Macroscopic Models of Chemotactic Movement

In the previous section we have established, departing from individual cell movement, the Patlak/Keller-Segel system (2.11)–(2.12) which describes the chemotactic movement of cells at the population scale. We now give properties of solutions of this system that we simplify as the *parabolic-elliptic* version, that is

$$\begin{cases} \frac{\partial}{\partial t}n(x,t) - \Delta n(x,t) + \mathrm{div}(n\chi\nabla c) = 0, & x \in \mathbb{R}^2, \\ -\Delta c(x,t) = n(x,t), \\ n(x,t=0) = n^0(x) \geq 0. \end{cases} \tag{2.17}$$

Notice that the solution of the equation for c is not well defined, and in fact we use the representation formula

$$\nabla c = -\lambda_d \int_{\mathbb{R}^d} \frac{x-y}{|x-y|^d} n(y,t)dy = -\lambda_d \frac{x}{|x|^d} \star n, \qquad \lambda_d = \frac{1}{d|B_d|}. \tag{2.18}$$

Let us mention that many variants and extensions of this system have been proposed. The reader can find useful references in [12, 26, 54, 69, 74, 92]. In this section, we only present some very specific and recent examples of models which prevent blow-up.

2.3.1 Elementary Properties

The Keller–Segel system admits several a priori estimates which reflects the basic modeling assumptions. The solution remains nonnegative and the total cell number is conserved

$$n(x, t) \geq 0, \qquad \int_{\mathbb{R}^d} n(x, t) dx := m^0 = \int_{\mathbb{R}^d} n^0(x) dx.$$

It also admits a dissipation principle for the free energy $\mathcal{E}(t)$ defined as

$$\begin{cases} \mathcal{E}(t) = \int_{\mathbb{R}^d} [n \ln(n) - \frac{\chi}{2}|\nabla c|^2] \, dx, \\ \frac{d}{dt}\mathcal{E}(t) = - \int_{\mathbb{R}^d} n|\nabla \ln(n) - \chi \nabla c|^2 \, dx \leq 0. \end{cases} \qquad (2.19)$$

The main property of this free energy $\mathcal{E}$ is that it is composed of two terms; the entropy $\int_{\mathbb{R}^d} n \ln(n)$ (which is essentially positive because small values of n do not count in practice) and the (negative) potential energy $-\frac{\chi}{2}\int_{\mathbb{R}^d} |\nabla c|^2$. These different signs allow for a competition between the dissipative (diffusion) term and the drift term.

Proof To derive this free energy principle, we write the equation on n in (2.17) as

$$\frac{\partial}{\partial t} n(x, t) = \text{div}[n(\nabla \ln(n) - \chi \nabla c)].$$

Therefore, multiplying by $\ln(n) - \chi c$ and integrating by parts, we obtain

$$\int_{\mathbb{R}^d} (\ln(n) - \chi c) \frac{\partial}{\partial t} n(x, t) = - \int_{\mathbb{R}^d} n|\nabla \ln(n) - \chi \nabla c|^2.$$

It remains to compute

$$\int_{\mathbb{R}^d} c \frac{\partial}{\partial t} n(x, t) \, dx = - \int_{\mathbb{R}^d} c \frac{\partial}{\partial t} \Delta c \, dx = \int_{\mathbb{R}^d} [\nabla c \frac{\partial}{\partial t} \nabla c \, dx$$

$$= \frac{1}{2} \int_{\mathbb{R}^d} \frac{\partial}{\partial t} |\nabla c|^2 \, dx.$$

Remark 2.1 When considering a more complete equation for c

$$\tau \frac{\partial}{\partial t} c - \Delta c + \alpha c = n,$$

the free energy and energy dissipation are given by the expressions

$$\begin{cases} \mathcal{E}(t) = \int_{\mathbb{R}^d} \left[n \ln(n) - \chi nc + \frac{\chi}{2}[|\nabla c|^2 + \alpha c^2] \right] dx, \\ \frac{d}{dt}\mathcal{E}(t) = - \int_{\mathbb{R}^d} n|\nabla \ln(n) - \chi \nabla c|^2 \, dx - \frac{\chi \tau}{\alpha} \int_{\mathbb{R}^d} |\frac{\partial c}{\partial t}|^2 dx \leq 0. \end{cases}$$

2.3.2 Blow-Up in the Keller–Segel System

We recall the following results, see for instance [13, 74] and the references therein,

Theorem 2.2 *In dimension $d = 2$, for the Patlak/Keller–Segel system with initial data satisfying $\int_{\mathbb{R}^2} n^0[1 + |x|^2 + |\log(n^0)|]dx < \infty$, we have*

(i) for $\|n^0\|_{L^1(R^2)} < \frac{8\pi}{\chi}$ there are smooth solutions of (2.17),
(ii) for $\|n^0\|_{L^1(R^2)} > \frac{8\pi}{\chi}$ solutions blow-up in finite time (as a singular measure),
(iii) for radially symmetric solutions, blow-up means

$$n(t) \approx \frac{8\pi}{\chi}\delta(x = 0) + Remainder.$$

An very important literature is devoted to this blow-up phenomena. Among them, let us mention the recent blow-up result in the parabolic-parabolic case [68].

Proof We do not not give a complete proof but we present an elementary calculation which explains why the solutions have to blow-up. This calculation is based on the second moment of the solution

$$m_2(t) := \int_{\mathbb{R}^2} \frac{|x|^2}{2}n(x, t)dx.$$

We have, from the first equation in (2.17),

$$\frac{d}{dt}m_2(t) = \int_{\mathbb{R}^2} \frac{|x|^2}{2}[\Delta n - \operatorname{div}(n\chi\nabla c)]dx$$

$$= \int_{\mathbb{R}^2}[2n + \chi nx \cdot \nabla c]dx$$

$$= 2m^0 - \chi\lambda_2 \int_{\mathbb{R}^2 \times \mathbb{R}^2} n(x, t)n(y, t)\frac{x \cdot (x-y)}{|x-y|^2}$$

$$= 2m^0 - \frac{\chi\lambda_2}{2} \int_{\mathbb{R}^2 \times \mathbb{R}^2} n(x, t)n(y, t)\frac{(x-y) \cdot (x-y)}{|x-y|^2}$$

(this last equality just follows by a symmetry argument, interchanging x and y in the integral). This yields finally,

$$\frac{d}{dt}m_2(t) = 2m^0(1 - \frac{\chi}{8\pi}m^0).$$

Therefore if we have $m^0 > 8\pi/\chi$, we arrive at the conclusion that $m_2(t)$ should become negative in finite time which is impossible since n is nonnegative. Therefore the solution cannot be smooth enough to justify this calculation.

Both theoretical analysis and numerical simulations show that solutions exhibit Dirac mass singularities. These are the most common patterns exhibited by the Patlak/Keller–Segel system.

This blow-up phenomena is compatible with observations of the amoeba *Dyctiostelium discoideum* moving on a dish. Chemotaxis leads them to form highly concentrated patterns, which we can interpret as a final stage before they change their comportment and thus the model becomes wrong. At this stage, the cells form a three dimensional multicellular fruiting body which generates spores that can disperse. But for *E. coli*, and for many other types of experiments, such a final stage is not observed and Dirac masses are not a desirable representation of the observations. This means the Keller–Segel models misses some features.

2.3.3   Keller–Segel System with Prevention of Overcrowding

In order to circumvent the difficulty of blow-up in finite time, a limitation of the drift term can be imposed in order to take into account volume filling effects and quorum sensing, see [54, 80]. The system (2.17) can be modified, for instance, as

$$\begin{cases} \frac{\partial}{\partial t}n(x,t) - D\Delta n(x,t) + \text{div}(n\psi(n)\nabla c) = 0, & x \in \mathbb{R}^2, \ t \geq 0, \\ -\Delta c(x,t) + \alpha c(x,t) = n(x,t), \end{cases} \tag{2.20}$$

with $\alpha > 0$ a degradation rate, $\psi(n)$ a switch for n large, for instance $\psi(n) = e^{-n/n_s}$. Another example is the logistic form $\psi(n) = n_s - n$. Then solutions remain bounded, $n \leq n_s$, when this is true for the initial data, thanks to the maximum principle. A remarkable feature of the system is its ability to form patterns. The paper [80] presents a complete analysis of the parameter range for instability and for the dynamics of patterns formed by this system.

Adapting [80], we can simply explain why patterns are formed. For any constant $\bar{n} > 0$ there is a steady state $n \equiv \bar{n}$, $S \equiv \bar{n}/\alpha$. This state is unstable when the inequality is satisfied:

$$\bar{n}\psi(\bar{n}) > D\alpha. \tag{2.21}$$

To see that, we look for a growing perturbation $n = \bar{n} + \delta e^{ikx}e^{\lambda t}$, $S = \frac{\bar{n}}{d} + \beta e^{ikx}e^{\lambda t}$. Inserting the expansion for δ, β small in the equations, gives

$$\begin{cases} \lambda\delta + D|k|^2\delta - \bar{n}\psi(\bar{n})\beta|k|^2 = 0, \\ (|k|^2 + \alpha)\beta = \delta, \end{cases}$$

that is also written

$$\lambda = -D|k|^2 + \bar{n}\psi(\bar{n})\frac{|k|^2}{|k|^2 + \alpha}.$$

We find a growing mode $\lambda > 0$ under the stated condition (2.21).

In one space dimension, steady state patterns can also be computed considering the steady states of (2.20). Assuming they decay fast enough at infinity, and considering the simpler elliptic case for the chemoattractant; they are solutions of

$$-Dn' + n\psi(n)c' = 0, \qquad -c'' = n.$$

With the function Q such that $Q'(n) = \frac{D}{n\psi(n)}$, we find $-Q(n)' + c' = 0$ that gives us

$$-c'' = -Q(n)'' = n.$$

This is a standard equation that can be studied and the range of parameters for existence can be uncovered, see [80].

Another effect which changes the blow-up conditions, and can possibly avoid it completely when q is large, is to use nonlinear diffusion and, then, the Keller–Segel system reads

$$\frac{\partial}{\partial t}n(x, t) - D\Delta n(x, t)^q + \mathrm{div}(n\nabla c) = 0$$

with a power $q > 1$. An important literature is also available for this case, see for instance [14, 27, 59] and the references therein.

2.3.4 The Flux Limited Keller–Segel System

As it can be derived from the kinetic equation for run and tumble, see Sect. 2.2.4, the Flux Limited Keller–Segel system is written,

$$\frac{\partial}{\partial t}n(x, t) + \mathrm{div}\left[n\,\chi\left(\frac{\partial c}{\partial t}, |\nabla c|\right)\nabla c\right] = 0, \tag{2.22}$$

where the nonlinear sensitivity χ has the form derived from the kinetic equation. We refer to [8, 35, 86, 87] for other variants of this system.

In [75], it is found that, with some coefficient A, and still assuming that the velocity set V is radially symmetric)

$$\chi\left(\frac{\partial c}{\partial t}, |\nabla c|\right) = \frac{A}{|\nabla c|}\int_V \frac{\xi_1 d\xi}{\Phi\left(\frac{\partial c}{\partial t} + \xi_1|\nabla c|\right)}.$$

The non-negativity of the chemotactic sensitivity χ is a consequence of the assumption that Φ is decreasing.

Notice that when $\nabla c = 0$, we have

$$\int_V \frac{\xi_1 d\xi}{\Phi\left(\frac{\partial c}{\partial t} + \xi_1|\nabla c|\right)} = \int_V \frac{\xi_1 d\xi}{\Phi\left(\frac{\partial c}{\partial t}\right)} = 0.$$

This avoids an indetermination of $\frac{\nabla c}{|\nabla c|}$ when $\nabla c = 0$.

However when using the stiff response limit at the kinetic scale, this indetermination arises, because χ is given by the ratio

$$\chi = \chi_0 \frac{\nabla c}{|\nabla c|},$$

which leads to a particularly subtle theory developed in [58].

2.3.5 Traveling Bands

It is commonly admitted that chemotaxis is one of the key ingredients triggering the formation of traveling bands (pulses) as observed in Adler's famous experiment for *E. Coli* (1966), [1]. We refer to [90] for a complete review of experimental assays.

Recently a mathematical and quantitative explanation has been developed in [86, 87], using the Flux-Limited Keller–Segel system with nutrient S in dimension $d = 1$

$$\begin{cases} \frac{\partial}{\partial t} n(x, t) - \Delta n(x, t) + \mathrm{div}[n(U_c + U_S)] = 0, \\[2mm] U_c = \chi_c \frac{\nabla c}{|\nabla c|}, \qquad U_S = \chi_S \frac{\nabla S}{|\nabla S|}, \\[2mm] \frac{\partial c}{\partial t} - D_c \Delta c + \alpha c = \beta n, \qquad \frac{\partial S}{\partial t} - D_S \Delta S = -\gamma n S. \end{cases}$$

Traveling waves are defined as solutions of the form $n(x - \sigma t) > 0$, $c(x - \sigma t)$, $S(x - \sigma t)$ for which $n(\pm\infty = 0)$. The parameter $\sigma \in \mathbb{R}$ is called the traveling speed and is due to the movement toward fresh nutrient S. If S is ignored, then standing pulses are observed in accordance with the experimental observations in [67].

In [86], in the case with stiff response, traveling pulses to this FLKS model are built analytically and they exhibit an asymmetric profile as it is observed experimentally. For a more general response function Φ, it is difficult to assert the existence of pulses. Another important and difficult extension is to assert the existence of traveling or standing bands (pulses) for the kinetic equation, we refer to [15, 26, 28].

2.3.6   Instabilities

Chemotaxis is a major phenomena which produces patterns as observed in nature. Parabolic models as the Patlak/Keller–Segel system have been widely used for such purposes, cf. [54, 70, 80] and the references therein. As mentioned before the limitation of the drift is a possible mechanism in this direction. Another direction is flux limitation as shown in [29].

There is only a limited literature on instabilities and pattern formation ability of solutions to the kinetic equations of bacterial chemotaxis. We refer to [76] where the stiffness parameter in the tumbling kernel with modulation along the path, that is δ small in (2.16), appears to be a bifurcation parameter.

2.4   Compressible Models of Tissue Growth

Models of tissue growth appear in the development of organisms, in tissue regeneration and also in tumor growth. These models serve not only to predict the evolution of cancers in medical treatments, using model based image analysis for example [84], but also to understand the biological and mechanical effects that are involved in the tissue growth.

The models under consideration contain several levels of complexity, both in terms of the biological and mechanical effects, and therefore in their mathematical description. We begin with a class of models that can be considered as "compressible fluids" that means that the pressure depends directly on the cell number density. Such models are reminiscent from a numerous literature, see for instance [11, 21, 34, 88]. Again our goal here is to describe the models from a mathematical point of view and the connections between different models.

2.4.1   A Simple Model with a Single Type of Cells

Solid tumors grow under the effect of cell proliferation limited by several factors. Space availability, and the pressure induced by higher cell population, appears to be the first cause of growth limitation by contact inhibition [11, 22, 83]. This can be included in the simplest models for a cell population density $n(x, t)$ where pressure generates both movement and growth limitation, leading to write

$$\begin{cases} \frac{\partial}{\partial t}n + \operatorname{div}(nv) = nG(p), & x \in \mathbb{R}^d, \ t \geq 0, \\ n(x, t = 0) = n^0(x) \geq 0, \\ v(x, t) = -\nabla p(x, t), & p(x, t) \equiv \Pi_\gamma\big(n(x, t)\big) := n(x, t)^\gamma, \quad \gamma > 1. \end{cases}$$
$$(2.23)$$

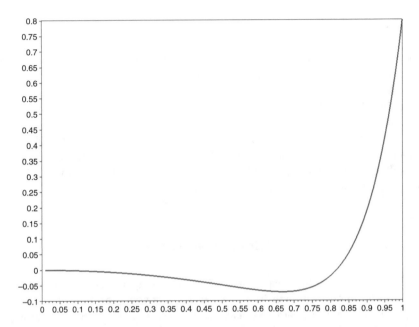

Fig. 2.1 A realistic pressure law $\Pi(n)$ should be decreasing for low densities n and increasing after, see for instance [34, 56] and is possible in the framework of the Cahn-Hilliard model, see Sect. 2.4.5. The power law we use here is a simplification compatible with the degenerate parabolicity of Eq. (2.23)

The rule $v(x, t) = -\nabla p(x, t)$ is a simplified version of Darcy's law expressing isotropic and homogeneous friction with the surrounding environment (fibers). This expression for the velocity field means that cells are only pushed by mechanical forces (variants are mentioned later). The particular choice for the law-of-state $\Pi_\gamma(n) := n^\gamma$ is made for simplicity, see considerations on this issue in [36] and Fig. 2.1. Finally the growth term, the right hand side in (2.23), is of Lotka-Volterra type, and takes into birth and death of cells. Because pressure generates contact inhibition, we assume that the C^1 function $G(\cdot)$ satisfies

$$G(0) = G_M > 0, \qquad G'(\cdot) < 0, \qquad G(P_h) = 0, \qquad \text{for some } G_M > 0, \ P_h > 0. \tag{2.24}$$

The name 'homeostatic pressure' has been proposed for value P_h ([83]) when cells stop proliferating. At this stage it might also be useful to mention that dimensions $d = 2$ is relevant for *in vitro* experiments on a dish and $d = 3$ is relevant both *in vitro* and *in vivo*.

This equation is a semi-linear version of the porous medium equation

$$\frac{\partial}{\partial t} n - \Delta n^m = 0, \qquad m > 1,$$

which has been widely studied, see the references in [91]. Therefore, from the standard theory of porous media equation, several properties and bounds can be derived, under appropriate assumptions on the initial data. However, as we see it further, the interaction between the growth term and the differential part yields new interesting features.

As well-known for the porous media equation, the equation on the pressure is extremely useful. Namely, we compute for later use

$$\frac{\partial}{\partial t} p - n\Pi'(n)\Delta p - |\nabla p|^2 = n\Pi'(n)G\big(p(x,t)\big). \tag{2.25}$$

Before we go further, we would like to insist that all the a priori bounds below are independent of the parameter γ.

2.4.1.1   Supersolution with Bounded Support

It is well known that a property of the porous medium equation is to describe solutions with compact support that expands. Therefore, for our purpose here, we do not bother with a bounded domain and associated boundary conditions. This feature is however relevant both for realistic models and numerics.

We also work on the equation written on the pressure (2.25)

Lemma 2.1 *There is a family of supersolutions under the form*

$$P_S(x,t) = B\big(R(t)^2 - |x|^2\big)_+,$$

for $B \geq \frac{G_M}{2d}$, $R(t) = R(0)e^{2Bt}$.

Choosing B and $R(0)$ allows a large choice of functions larger than any bounded initial data with bounded support. Therefore we have

Corollary 2.1 *For n^0 bounded and with bounded support, the solution of (2.23) has bounded support.*

Proof To prove Lemma 2.1, we compute

$$\frac{\partial P_S(x,t)}{\partial t} = 2BR\dot{R}\mathbb{1}_{\{|x|\leq R(t)\}} \geq 4B^2R^2\mathbb{1}_{\{|x|\leq R(t)\}} \geq |\nabla P_S|^2 = 4B^2|x|^2\mathbb{1}_{\{|x|\leq R(t)\}}$$

and, for some $\mu > 0$

$$-\Delta P_S = -\mu\delta_{\{|x|=R(t)\}} + 2Bd\mathbb{1}_{\{|x|\leq R(t)\}} \geq G_M\mathbb{1}_{\{|x|\leq R(t)\}}.$$

Since the Dirac mass on the sphere $\{|x| = R(t)\}$ is killed by the vanishing pressure in the term $P_S\Delta P_S$, we conclude that

$$-P_S\Delta P_S \geq P_S G_M \geq P_S G(P_S).$$

And we have proved that

$$\frac{\partial P_S(x,t)}{\partial t} - \gamma P_S \Delta P_S - |\nabla P_S|^2 \geq \gamma P_S G(P_S).$$

2.4.1.2   Existence of Solutions and a Priori Bounds

Now, we follow closely [77]. Because we are interested in the dependence on the parameter γ (and large values of it), we consider a family of initial data n_γ^0 such that for some constant K^0,

$$\int_{\mathbb{R}^d} n^0 dx \leq K^0, \qquad p^0 := \Pi(n^0) \leq P_h, \qquad \int_{\mathbb{R}^d} |\nabla n^0| dx \leq K^0. \qquad (2.26)$$

Proposition 2.1 *With assumptions (2.24)–(2.26), the unique solution of Eq. (2.23) satisfies $n(x,t) \geq 0$ and*

$$\int_{\mathbb{R}^d} n(x,t) dx \leq K^0 e^{G_M t}, \qquad \int_{\mathbb{R}^d} |\nabla n(x,t)| dx \leq K^0 e^{G_M t},$$

$$\int_0^T \int_{\mathbb{R}^d} |\nabla p(x,t)| dx dt \leq C(T, P_h, K^0),$$

$$p(x,t) \leq P_h, \qquad \int_{\mathbb{R}^d} p(x,t) dx \leq P_h^{(\gamma-1)/\gamma} K^0,$$

$$\int_0^T \int_{\mathbb{R}^d} |\nabla p(x,t)|^2 dx dt \leq \frac{1 + \gamma G_M T}{\gamma - 1} P_h^{(\gamma-1)/\gamma} K^0.$$

Proof The estimates for n are straightforward. For the TV bound, we just notice that, the equation for n can also be written

$$\frac{\partial}{\partial t} n - \Delta \Phi(n) = nG(p(x,t)), \qquad \text{with } \Phi'(n) = n\Pi_\gamma'(n).$$

Therefore, the equation for $w_i = \frac{\partial n(x,t)}{\partial x_i}$ is

$$\frac{\partial}{\partial t} w_i - \text{div}[\Phi'(n)\nabla w_i] = w_i G(p(x,t)) + nG'(p(x,t)) \frac{\partial p(x,t)}{\partial x_i},$$

and finally, since the sign of $\frac{\partial p(x,t)}{\partial x_i}$ is the same than the sign of $\frac{\partial n(x,t)}{\partial x_i}$,

$$\frac{\partial}{\partial t}|w_i| - \text{div}[\Phi'(n)\nabla|w_i|] = |w_i|G(p(x,t)) - n|G'(p(x,t))| \, |\frac{\partial p(x,t)}{\partial x_i}| \le |w_i|G_M.$$

After integration and use of the Gronwall lemma, this gives the L^1 estimate on the gradient of n and keeping the term with $|\frac{\partial p}{\partial x_i}|$ gives the bound on the gradient of p (see [77] for details).

The third line of bounds in Proposition 2.1 follows from the equation on the pressure (2.25). This equation is in the strong form, the maximum principle applies and gives the bound $p \le P_h$. It gives the L^1 control on p because

$$p = n^\gamma = nn^{\gamma-1} = np^{(\gamma-1)/\gamma} \le nP_h^{(\gamma-1)/\gamma},$$

and it remains to apply the L^1 control on n.

The L^2 estimate on the gradient is easier to see when identifying the pressure law, as $\Pi(n) = n^\gamma$ in (2.25), to find

$$\frac{\partial}{\partial t}p - \gamma p\Delta p - |\nabla p|^2 = \gamma pG(p). \tag{2.27}$$

Integrating by parts, we obtain, for $T > 0$,

$$\int_{\mathbb{R}^d}[p(x,T) - p^0(x)]dx + (\gamma-1)\int_0^T\int_{\mathbb{R}^d}|\nabla p|^2dxdt \le \gamma G_M\int_0^T\int_{\mathbb{R}^d}p(x,t)dxdt.$$

which, combined with the L^1 estimate for p gives the last inequality which concludes the estimates of Proposition 2.1.

The bounds in Proposition 2.1 are fine to ensure compactness in space. It remains to prove estimates implying time compactness. An easy way is to notice that under the assumption that n^0 is a sub-solution, that is

$$-\text{div}(n^0\nabla\Pi(n^0)) \le n^0G(p^0(x)),$$

we have $\frac{\partial n^0}{\partial t} \ge 0$. We may apply the same argument as for space derivatives and $w = \frac{\partial n}{\partial t}$ satisfies

$$\frac{\partial}{\partial t}w - \text{div}[\Phi'(n)\nabla w] = wG(p(x,t)) + nG'(p(x,t))\gamma n^{\gamma-1}w,$$

an equation which gives us the property

$$\frac{\partial}{\partial t} n^0 \geq 0 \implies \frac{\partial n}{\partial t} \geq 0. \tag{2.28}$$

Because $\frac{\partial}{\partial t} n^0 - \mathrm{div}\left(n^0 \nabla p^0\right) = n^0 G(p^0)$, the assumption $\frac{\partial}{\partial t} n^0 \geq 0$ means that n^0 is a sub-solution of the stationary equation.

This property (2.28) is very strong and shows one limitation of the model at hand. It is incompatible with the observations that the cell population decreases in the center of the tumor, the necrotic core. This effect can be obtained when the nutrients are included in the equation, see (2.31) below.

In this situation, which we call 'well prepared initial data', we conclude

$$\frac{d}{dt} \int_{\mathbb{R}^d} |w(x,t)| dx \leq G_M \int_{\mathbb{R}^d} |w(x,t)| dx,$$

and thus

$$\int_{\mathbb{R}^d} \left| \frac{\partial}{\partial t} n(x,t) \right| dx \leq e^{G_M t} \int_{\mathbb{R}^d} \left| \mathrm{div}\left(n^0 \nabla \Pi(n^0)\right) + n^0 G\left(p^0(x)\right) \right| dx. \tag{2.29}$$

2.4.1.3 A Variant of Aronson-Bénilan Estimate

It is possible to improve these estimates and avoid the restrictive assumption that the initial data is a sub-solution of the stationary equation. We recall from [77] the

Proposition 2.2 *Assuming that* $r_G := \min_{p \in [0, p_h]} \left(G(p) - pG'(p)\right) > 0$, *the estimates hold, for all* $t > 0$,

$$\Delta p(t) + G(p(t)) \geq -r_G e^{-(\gamma-1)r_G t} / (1 - e^{-(\gamma-1)r_G t}). \tag{2.30}$$

$$\frac{\partial}{\partial t} p(x,t) \geq -\gamma\, r_G\, p(x,t) \frac{e^{-\gamma r_G t}}{1 - e^{-\gamma r_G t}}, \qquad \frac{\partial}{\partial t} n(x,t) \geq -r_G\, n(x,t) \frac{e^{-\gamma r_G t}}{1 - e^{-\gamma r_G t}}.$$

These inequalities allow for a fast transition at $t = 0$ (the right hand side is singular then). They were discovered in [6] for the porous media equation and improved in [40]. The improvement from [77] is to include the growth term G in the estimate 2.30 on Δp. They are much stronger than those in (2.28) because they do not assume any further assumption on the initial data than those in Proposition 2.1. A remarkable feature here, is that the semi-linear source term improves the usual inequalities for porous medium equations, which are recovered for $r_G \to 0$. Indeed it implies an exponential decay rather than algebraic.

2.4.2   Single Cell Type Population Model with Nutrient

Many additional effects are used in more realistic models of tumor growth. A possible additional improvement is to take into account nutrients. Then we arrive to the model, also treated in details in [77],

$$\begin{cases} \partial_t n - \operatorname{div}(n\nabla p) = n\ \Phi(p,c), \\ \partial_t c - \Delta c = -n\ \Psi(p,c), \\ c(x,t) = c_B > 0 \qquad \text{as } |x| \to \infty, \end{cases} \tag{2.31}$$

where c denotes the density of nutrients, and c_B the far field supply of nutrients (from blood vessels). The coupling functions Φ, Ψ are assumed to be smooth and to satisfy the intuitive hypotheses

$$\begin{cases} \partial_p \Phi < 0, & \partial_c \Phi \geq 0, & \Phi(P_h, c_B) = 0, \\ \partial_p \Psi \leq 0, & \partial_c \Psi \geq 0, & \Psi(p,0) = 0. \end{cases} \tag{2.32}$$

Variants are possible; for instance, we could assume that several nutrients (oxygen, glucose) are released continuously from a vasculature or an other source [36], and this can play an important role in cancer development.

As far as estimates are concerned, the solutions are also controlled with BV estimates under the following assumptions on the initial data, in addition to (2.26). We assume that for some c^0 such that $c_B - c^0 \in L^1_+(\mathbb{R}^d)$,

$$\begin{cases} 0 \leq c^0 < c_B, & \|(c^0)_{x_i}\|_{L^1(\mathbb{R}^d)} \leq C, \quad i = 1,\ldots,d, \\ \|\operatorname{div}(n^0 \nabla p^0) + n^0\ \Phi(p^0, c^0)\|_{L^1(\mathbb{R}^d)} \leq C, & \|\Delta c^0 - n^0\ \Psi(p^0, c^0)\|_{L^1(\mathbb{R}^d)} \leq C. \end{cases} \tag{2.33}$$

Let us however point out that the analog of the Aronson-Bénilan type estimate stated in Proposition 2.2 are not know for the model with nutrient.

Another remarkable effect, is that solutions may be very different and undergo instabilities due to local nutrient depletion, see [78]. This is not possible with the simpler model with a single equation.

2.4.3   Models with Two Cell Types

It is more realistic to take into account not only tumor cells but also other types of cells. These can be surrounding healthy cells, or quiescent (non-proliferative) tumor cells, and possibly necrotic cells which die by lack of nutrients. This leads to couple equations as in (2.23) with a global pressure resulting from all the cells. Considering only two types of cells for simplicity, we arrive at

$$
\begin{cases}
\partial_t n_1 - \text{div}[n_1 \nabla p] = n_1 F_1(p) + n_2 G_1(p), & x \in \mathbb{R}^d, \ t \geq 0, \\[2mm]
\partial_t n_2 - \text{div}[n_2 \nabla p] = n_1 F_2(p) + n_2 G_2(p),
\end{cases}
\tag{2.34}
$$

with

$$
n := n_1 + n_2, \qquad p = n^\gamma, \quad \gamma > 1.
\tag{2.35}
$$

The existence theory is much more complicated for such systems and we refer to the papers [11] for strong solutions to the non-degenerate case in a bounded domain when $n > 0$, to [31] for the degenerate in one space dimension and to [52] for the multidimensional case with a restriction on the growth functions.

Notice that the concentrations $c_i = \frac{n_i}{n_1 + n_2}$ satisfy the equation

$$
\partial_t c_i - \nabla c_i . \nabla p = c_1 F_i(p) + c_2 G_i(p) - c_i R,
$$

$$
R = c_1[F_1(p) + F_2(p)] + c_2[G_1(p) + G_2(p)].
$$

Therefore we find

$$
\partial_t (c_1 c_2) - \nabla(c_1 c_2) \nabla p = RHS,
$$

and thus

$$
\partial_t (n c_1 c_2) - \text{div}\big[n c_1 c_2 \nabla p\big] = n RHS,
\tag{2.36}
$$

an equation which can be compared with the two phase Cahn–Hilliard model in Sect. 2.4.5. It indicates a degenerate mobility when $c_1 = 0$ or $c_2 = 0$.

The possibility to write a closed equation as (2.36) is related to the so-called segregation property which we discuss in Sect. 2.5.3.

2.4.4 Two Cell Type Model with Different Mobilities

More realistic is also that healthy cells and tumor cells do not react to the pressure field with the same mobilities. This can be explained by adhesion proteins across the membrane of the cells which are more numerous or effective in healthy tissues than for tumor cells. The problem becomes far more delicate and many instabilities can occur, see [56, 63], generating dendritic patterns as in the Safman–Taylor instability.

The system with two cell type reads

$$\begin{cases} \partial_t n_1 - \text{div}[n_1\mu_1\nabla p] = n_1 F_1(p) + n_2 G_1(p), & x \in \mathbb{R}^d, \ t \geq 0, \\ \partial_t n_2 - \text{div}[n_2\mu_2\nabla p] = n_1 F_2(p) + n_2 G_2(p), \end{cases}$$

still with a law of state given for simplicity as

$$n := n_1 + n_2, \qquad p = n^\gamma, \quad \gamma > 1.$$

The mobilities factor μ_i can also depend on cell densities. This occurs in porous media equation and oil recovery problems, similar models occur [3–5, 30].

2.4.5 Surface Tension and the Degenerate Cahn–Hilliard Model

A widely used approach to describe the dynamics of two cells types takes into account surface tension between the phases (see Sect. 2.5.4) and relies on the degenrate Cahn–Hilliard model. A phase field $\phi(x, t)$ is used which describes the concentration of cells. For instance $\phi = 0$ represents healthy cells and $\phi = 1$ represents tumor cells. Following [2, 34, 48, 64], we write a continuity equation following Darcy's law

$$\frac{\partial}{\partial t}\phi - \text{div}[\mu(\phi)\nabla p] = 0,$$

where the mobility $\mu(\cdot)$ takes into account the degeneracy at phase saturation, for instance

$$\mu(\phi) = \phi(1 - \phi)^2,$$

a relation which garantees that $0 \leq \phi \leq 1$.

The main difference with previous models occurs in the 'pressure' term. It takes into account energy carried by the diffuse interface of size γ separating the phases

$$p = -\gamma\Delta\phi - F'(\phi)$$

and the stability of the phases $\phi = 0$ and $\phi = 1$ results in expressions for the interaction potential of the type (see Fig. 2.1)

$$F'(\phi) = \frac{\phi^2(\phi - \phi^*)}{1 - \phi} \qquad \text{or} \qquad F(\phi) = -\frac{1}{2}\phi\ln\phi - (1 - \phi)\ln(1 - \phi) + (\phi - \frac{1}{2})^2.$$

In the original model of Cahn–Hilliard for material science, where $\mu \equiv 1$, the potential is double well. Here the phase $\phi = 0$ is metastable, see [23] and [2]. For this reason, the model is able to represent phase separation which is often observed in living tissues. Cells of the same type have a tendency to aggregate in patches.

The mechanistic consistency of the model is expressed by two relations, the energy and the entropy. For the energy, we define

$$E_{CH}(t) = \int \left[\frac{\gamma}{2} |\nabla \phi|^2 - F(\phi) \right] dx.$$

We compute

$$\frac{dE_{CH}(t)}{dt} = - \int \frac{\partial \phi}{\partial t} [\gamma \Delta \phi + F'(\phi)] dx$$

$$= \int \mu(\phi) \nabla p \nabla [\gamma \Delta \phi + F'(\phi)] dx = - \int \mu(\phi) |\nabla p|^2 \leq 0.$$

Another physical quantity is the entropy relation. Define the mapping $S : [0, \infty) \mapsto [0, \infty)$ using

$$S''(\phi) = \frac{1}{\mu(\phi)}, \qquad S(0) = S'(0) = 0.$$

The entropy functional is defined as

$$\Phi[\phi(t)] = \int S\big(\phi(x, t)\big) dx$$

It is useful to keep in mind that the entropy functional behaves as follows in the biophysical case $\mu(n) = n(1 - n)^2$

$$S(\phi) = \phi \ln(\phi), \ \phi \approx 0^+, \qquad S(\phi) = -\log(1 - \phi), \ \phi \approx 1^-.$$

The entropy relation writes

$$\frac{d\Phi(t)}{dt} = - \int \left[\gamma |\Delta \phi|^2 + F''(\phi) |\nabla \phi|^2 \right].$$

Mathematical analysis of this type of equation is very active presently, see for instance [2, 30, 47] and the references therein.

2.5 Incompressible Models of Tissue Growth

Incompressible models suppose that the tissue is saturated by cells, liquids or more generally phases. They have the great advantage to avoid defining a pressure law because the pressure is interpreted as a Lagrange multiplier for the incompressibility constraint. For that reason these models are mostly selected in practical use of cancer models, realistic simulations and software development [37–39, 46, 64, 85].

The outcome is that free boundary problems occur which are of mathematical interest. The well-posedness analysis is interesting in itself but here we do not touch this issue. We just describe the equations in parallel to the compressible models of Sect. 2.4.

2.5.1 Single Cell Type Free Boundary Problem

The one cell type incompressible model, which corresponds to the compressible one of Sect. 2.4.1 is the following Hele-Shaw equation of fluid mechanics. It is a standard free boundary problem. The tumor occupies a domain $\Omega(t)$. The free boundary $\partial\Omega(t)$ of the domain $\Omega(t)$ is moving with the velocity (in fact the only normal derivative is needed)

$$v_\infty(x, t) = -\nabla p_\infty(x, t) \tag{2.37}$$

where the pressure field is computed thanks to the equation

$$\begin{cases} -\Delta p_\infty = G(p_\infty), & x \in \Omega(t), \\ p_\infty = 0 & \text{on } \partial\Omega(t). \end{cases} \tag{2.38}$$

In order to define this dynamic, some smoothness of the free boundary is necessary (on needs to solve the Dirichlet problem and define normal derivative at least). Such a property has been widely studied, see [45, 66] and the references therein. An alternative is to set this problem in the general framework of viscosity solutions with a correct viscosity condition on the interface, see [61]. Surface tension may also be included [45, 46], then the Dirichlet boundary condition has to be changed to $p_\infty = a\kappa(x, t)$ on $\partial\Omega(t)$, with a a parameter and κ the mean curvature of $\partial\Omega(t)$.

As we mentioned earlier, the biophysical modeling [22, 83], gives growth terms G that depend on p, and not on n, which closes very nicely Eq. (2.38).

It is not intuitive why this geometric free boundary problem is related to Eq. (2.23) and we discuss that in Sect. 2.6.

2.5.2 Single Cell Type Model with Nutrient and Free Boundary

When considering the effect of nutrients, we arrive to a simple variant of the geometric motion that we state for completeness. With the notations of Sect. 2.4.2, we find that the equation for the pressure (2.38) is replaced by

$$
\begin{cases}
-\Delta p_\infty = \Phi(p_\infty, c_\infty), & x \in \Omega(t), \\
p_\infty = 0 & \text{on } \partial\Omega(t).
\end{cases}
\tag{2.39}
$$

coupled with the equation for the nutrient

$$
\begin{cases}
\partial_t n_\infty - \text{div}(n_\infty \nabla p_\infty) = n_\infty \, \Phi(p_\infty, c_\infty), \\
\partial_t c_\infty - \Delta c_\infty = -n_\infty \, \Psi(p_\infty, c_\infty), \\
c_\infty(x, t) = c_B > 0 & \text{as } |x| \to \infty,
\end{cases}
\tag{2.40}
$$

2.5.3 Two Cell Types Incompressible Model

The case with two cell types gives another and maybe more intuitive example of the meaning of incompressible models. Using the equations of Sect. 2.4.3, the incompressible case is written as

$$
\begin{cases}
\partial_t n_1 - \text{div}[n_1 \nabla p] = n_1 F_1(p) + n_2 G_1(p), & x \in \mathbb{R}^d, \ t \geq 0, \\
\partial_t n_2 - \text{div}[n_2 \nabla p] = n_1 F_2(p) + n_2 G_2(p),
\end{cases}
\tag{2.41}
$$

and the pressure $p(x, t)$ is now a Lagrange multiplier for the constraint

$$
n := n_1 + n_2 = 1.
\tag{2.42}
$$

In other word, adding the two equations, we find

$$
-\Delta p = n_1[F_1(p) + F_2(p)] + n_2[G_1(p) + G_2(p)].
$$

This corresponds to the equation $-\Delta p_\infty = G(p_\infty)$ when $\Omega(t) = \mathbb{R}^d$ in the free boundary problem (2.38).

Let us recall that, in (2.38) the tumor boundary is determined by $\partial\Omega(t)$, while in the two phase model we expect that the tumor is $\{n_1 > 1/2\}$. In order to complete the analogy, the segregation property [11, 31] is desirable

$$
n_1(x, t) \, n_2(x, t) = 0.
$$

This property follows formally from the equation for $n_1(x, t) n_2(x, t) = 0$

$$\partial_t (n_1 n_2) - \text{div}[n_1 n_2 \nabla p] - n_1 n_2 \Delta p = n_1 n_2 [F_1(p) + G_2(p)] + n_1^2 F_2(p) + n_2^2 G_1(p).$$

under the condition $F_2(p) \equiv 0$, $G_1(p) \equiv 0$.

2.5.4 Multiphase Models

Continuing on the fluid mechanical view of a tissue, the formalism of multiphase fluids can be used in the present context [24, 25, 81, 82] in order to represent the complexity of cell surrounding.

In the simplest possible view that we retain here for simplicity,

- only two phases are considered: the tumor cells and a 'liquid' (a generic name for extracellular fluid) with local volume ratio $\phi_T(x, t) \geq 0$, $\phi_L(x, t) \geq 0$,
- the saturation regime (no void, incompressible) is written $\phi_T + \phi_L = 1$,
- the matter, tumor cells and liquid, flows with velocity v_T, v_L.

One writes usual Navier–Stokes equations for the different phases and their momentum, leading to the system describing mass exchange and momentum exchange

$$
\begin{cases}
\frac{\partial}{\partial t} \phi_T + \text{div}(\phi_T v_T) = \Gamma_T, \\
\frac{\partial}{\partial t} \phi_L + \text{div}(\phi_L v_L) = \Gamma_L, \\
\rho_T \phi_T [\frac{\partial}{\partial t} v_T + v_T . \nabla v_T] = \text{div} . \mathcal{T}_T + m_T, \\
\rho_L \phi_L [\frac{\partial}{\partial t} v_L + v_L . \nabla v_L] = \text{div} . \mathcal{T}_L + m_L, \\
\phi_T + \phi_L = 1.
\end{cases}
\tag{2.43}
$$

• The cell and liquid densities ρ_T and ρ_L (two constants) are usually assumed to be equal, which means that the mass balance $\rho_T \Gamma_T + \rho_L \Gamma_L = 0$ reads

$$\Gamma_T + \Gamma_L = 0.$$

As a consequence, we obtain the incompressibility relation

$$\text{div}(\phi_T v_T + (1 - \phi_T)v_L) = 0. \tag{2.44}$$

To fix ideas, a possible expression for this growth term is Fisher type (before necrosis occurs)

$$-\Gamma_L = \Gamma_T = G \, \phi_T \, \phi_L = G \, \phi_T \, (1 - \phi_T). \tag{2.45}$$

This expression can be compared to the term $nG(p)$ which we have used so far.

- The interaction forces, m_T, m_L, are due to interfacial pressure P and friction terms (friction between phases and for tumor cells an adhesion on the extracellular matrix)

$$m_T = P\nabla\phi_T + k_1\phi_T\phi_L(v_L - v_T) - k_2 v_T, \quad m_L = P\nabla\phi_L + k_1\phi_T\phi_L(v_T - v_L),$$
$$(2.46)$$

  with a dependency $k_2(\phi_T)$

- The stress tensors $\mathcal{T}_{T,L}$ represent internal forces to each phase. They should be defined in a closed form and there are many possible choices. It helps at this stage to neglect the acceleration terms $\frac{\partial}{\partial t}v + v.\nabla v$. Then the equations for the momentum are simplified to

$$\begin{cases} 0 = \text{div}.\mathcal{T}_T + m_T, \\ 0 = \text{div}.\mathcal{T}_L + m_L, \end{cases} \qquad (2.47)$$

To show the power off the formalism, e.g., [82], we give possible choices for the stress tensors.

2.5.4.1 The One Phase Closure

The simplest choice is to neglect interactions between the phases, that is $k_1 = 0$, $P = 0$. To close the system one takes $P_T = \Pi(\phi_T)$, $\mathcal{T}_T = -P_T\mathbf{I}$. One ends up with a very simplified system which can be written

$$\begin{cases} \frac{\partial}{\partial t}\phi_T + \text{div}(\phi_T v_T) = \Gamma_T = \phi_T(1 - \phi_T)G, \\ k_2 v_T = -\nabla\Pi(\phi_T) \qquad \text{(Darcy's law).} \end{cases} \qquad (2.48)$$

This is the simplest model that has been presented in Sect. 2.4.1.

However, the approach still gives some information about the dynamics of the liquid phase which is reduced to the incompressibility relation (2.44).

2.5.4.2 The Darcy/Stokes Closure

In the stress tensor, we retain only the pressure term for the cells and only the viscosity terms for the liquid. This means that the equations take the form

$$\mathcal{T}_T = -P_T\mathbf{I}, \quad P_T = \Pi(\phi_T), \qquad \text{(Darcy's law)}$$

$$\mathcal{T}_L = -\nabla P_L + \Delta v_L, \qquad \text{(Stokes equation)}$$

Here P_L is the Lagrange multiplier for the incompressibility relation (2.44).

2.5.4.3 The Single Pressure Closure

One may also choose to keep a single pressure $P(x,t)$, and again to retain the isotropic tensor $\mathcal{T}_{T,L} = -P_{T,L}\,\mathbb{I}$, and set

$$P_T = \phi_T\,P, \qquad\qquad P_L = \phi_L\,P.$$

This allows to solve for v_T and v_L in terms of P; first we add the two momentum equations in (2.47) and find $m_T + m_L = \nabla P$. We find

$$v_T = -\frac{1}{k_2}\nabla P, \qquad\qquad v_L = v_T - \frac{1}{k_1\phi_T}\nabla P.$$

This finally leads to the system in closed form

$$\begin{cases} \frac{\partial}{\partial t}\phi_T + \mathrm{div}\big(\phi_T v_T\big) = \Gamma_T, \\[2mm] v_T = -K_2(\phi_T)\nabla P(x,t), \\[2mm] \mathrm{div}[K_1(\phi_T)\nabla P] = 0. \end{cases} \qquad (2.49)$$

x,t This is also a standard two phase flow problem (in particular it is widely used for oil/gas mixtures in oil recovery). Our understanding is very little because it shares hyperbolic nature (having in mind that the pressure is smooth) and the parabolic nature of equations derived from Darcy's law.

To conclude this case, let us explain the a priori estimates $0 \le \phi_T \le 1$ directly from (2.49). Because Γ_T vanishes when $\phi_T = 0$, see (2.45), we get $-\phi_T \ge 0$. For the other inequality, we have to write the equation as

$$\frac{\partial}{\partial t}\phi_T - \mathrm{div}\Big(\phi_T\frac{K_2}{K_1}K_1\nabla P\Big) = \Gamma_T,$$

$$\frac{\partial}{\partial t}\phi_T + \nabla\left(\phi_T\frac{K_2(\phi_T)}{K_1(\phi_T)}\right).K_1\nabla P = \Gamma_T.$$

Solutions satisfy $\phi_T \le 1$ because it is in strong form and Γ_T vanishes for $\phi_T = 1$. These expressions show that these models can be classified as nonlinear hyperbolic conservation laws.

2.6 The Incompressible Limit and Stiff Pressure Law

As long as cells are well separated, the pressure forces are negligible. When the population density increases, there is a maximum possible compaction which cannot be exceeded. The transition is however very stiff. To represent this effect in compressible models, the simplest formalism is to consider the limit as $\gamma \to \infty$

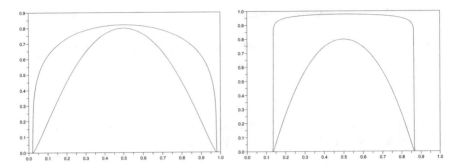

Fig. 2.2 *Effect of γ large.* A solution to the mechanical model (2.23) in one dimension with $G(p) = 5. * (1 - p)$. Left: $\gamma = 5$. Right: $\gamma = 40$. The upper line is n; the bottom line is p (scale enlarged for visibility). Notice that the density scales are not the same in the two figures. The initial data is taken with compact support and the solution is displayed for a time large enough (see Fig. 2.3 below for an intermediate regime). Reproduced from [77]

in the equation of state (2.23), and which we call the *stiff pressure asymptotic*. This limit provides a derivation of incompressible equations from compressible equations. The Fig. 2.2 illustrates the behavior of solutions for large values of γ in Eq. (2.23).

2.6.1 Single Cell Type Model, Incompressible Limit

We consider the derivation of the geometric Hele-Shaw free boundary problem (2.5.1) departing from (2.23). We first establish the weak formulation of the problem, including the possible singularities shown in Fig. 2.3. The difficult step is to establish the so-called complementary relation. Then, we show that for a nice initial data $n^0 = \mathbb{1}_{\{\Omega^0\}}$, then the geometric form is equivalent to the weak form.

In order to indicate the dependency on $\gamma \to \infty$, now denote by n_γ, p_γ the solutions of (2.23) as built in Sect. 2.4.1.2.

2.6.1.1 Weak Formulation of the Hele-Shaw Problem

Besides the free boundary formulation, there is also a weak formulation of the limit $\gamma \to \infty$ in Eq. (2.23). This limit gives a more general setting allowing a 'pretumor zone' where healthy and tumor cells are present in a mixed state. This weak formulation was derived in [77] and leads to the equation

$$\begin{cases} \frac{\partial}{\partial t} n_\infty - \text{div}(n_\infty \nabla p_\infty) = n_\infty G(p_\infty(x, t)), & x \in \mathbb{R}^d, \ t \geq 0, \\ n_\infty(x, t = 0) = n_\infty^0(x) \geq 0, \\ p_\infty(1 - n_\infty) = 0, & 0 \leq n_\infty \leq 1. \end{cases} \qquad (2.50)$$

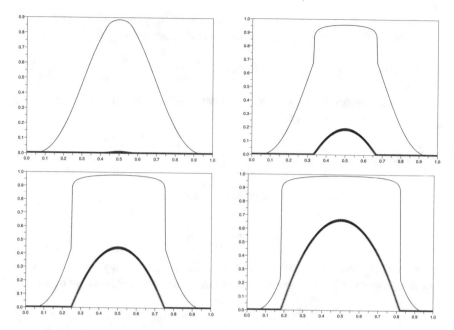

Fig. 2.3 *Cell density and pressure carry different informations.* Here $\gamma = 40$ and the initial data n is less than 1. The solution is displayed at four different times. It shows how the smooth part of n strictly less than 1 is growing with $p = 0$ (figure on the left). When n reaches the value 1, the pressure becomes positive, increases and creates a moving front that delimitates the growing domain where $n \approx 1$. Thin line is n and thick line is p as functions of x. Reproduced from [77]

In other words, when $n_\infty < 1$ then $p_\infty = 0$. Consequently, n_∞ and p_∞ are so weakly related that their dynamics can be somewhat independent. Nevertheless, a remarkable property is that the weak solution of (2.50) is unique (see [77]).

To present the result, we now insert the index γ to the notations n and p for the solutions of (2.23). The following result holds

Theorem 2.3 (Hele-Shaw Limit, [77]) *With the assumptions of Proposition 2.1, as $\gamma \to \infty$, we have*

$$n_\gamma \to n_\infty \leq 1, \qquad p \to p_\infty \leq P_h \qquad a.e. \text{ in } \mathbb{R}^d \times (0, \infty),$$

$$\nabla p_\gamma \rightharpoonup \nabla p_\infty \qquad \text{in } L^2\big(\mathbb{R}^d \times (0, T)\big) - weak, \qquad \forall T > 0,$$

$$\frac{\partial}{\partial t} n_\infty \geq 0, \qquad \frac{\partial}{\partial t} p_\infty \geq 0.$$

The limit of Eq. (2.23) is Eq. (2.50).

Notice that, from the BV (bounded variation) properties of n_γ and p_γ in Proposition 2.1, we derive strong compactness. We also conclude that

$$n_\infty \in L^\infty\big((0, T); L^1 \cap L^\infty(\mathbb{R}^d)\big), \quad p_\infty \in L^\infty\big((0, T) \times \mathbb{R}^d\big) \cap L^1\big((0, T) \times \mathbb{R}^d\big)$$

and that, as measures although we use the notation of L^1 functions, $|\nabla n_\infty(x, t)|$ and $|\nabla p_\infty(x, t)|$ are bounded with

$$\int_{\mathbb{R}^d} |\nabla n_\infty(x, t)|dx \le K^0 e^{GMt}, \quad \int_0^T \int_{\mathbb{R}^d} |\nabla p_\infty(x, t)|dxdt \le C(T, P_h, K^0).$$

The other results follow immediately. For example, because

$$n_\gamma p_\gamma = n^{\gamma+1} = p_\gamma^{\frac{\gamma+1}{\gamma}},$$

and passing to the strong limits, we find in the limit the relation $p_\infty(1 - n_\infty) = 0$. Another property follows immediately from the same argument; because $n_\gamma \nabla p_\gamma = \nabla p_\gamma^{\frac{\gamma+1}{\gamma}}$, we find the relation

$$n_\infty \nabla p_\infty = p_\infty.$$

In other words, the equation on n_∞, in (2.50), can also be written

$$\frac{\partial}{\partial t} n_\infty - \Delta p_\infty = n_\infty G\big(p_\infty(x, t)\big).$$

This is the form used in [77] to prove uniqueness of weak solutions.

2.6.1.2   The Complementary Relation

A more difficult result is the derivation of the 'complementary relation', (2.51) below, which is equivalent to the strong convergence of ∇p_γ.

Theorem 2.4 (Complementary Relation) *Additionally to Theorem 2.3, one also has*

$$\nabla p_\gamma \to \nabla p_\infty \qquad \text{in } L^2_{\text{loc}}(\mathbb{R}^d \times (0, \infty)) - strong,$$

The 'complementary relation' also holds in $\mathcal{D}(\mathbb{R}^d \times (0, \infty))$

$$p_\infty \big(\Delta p_\infty + G(p_\infty)\big) = 0. \tag{2.51}$$

The *complementary relation* (2.51) is not an obstacle problem (a sign is incompatible) and the solution is not unique. It is a weak version of Eq. (2.38) with

$$\Omega(t) = \{ \, p_\infty(x, t) > 0 \, \}, \tag{2.52}$$

a set which evolution cannot be deduced from (2.51) alone, but follows from the weak formulation (2.50).

The meaning, in distributions, of (2.51) is that for all smooth test functions φ with compact support, it holds

$$\int_{\mathbb{R}^d \times (0,\infty)} \varphi(x, t) \left[-|\nabla p_\infty|^2 + p_\infty G(p_\infty) \right] - \int_{\mathbb{R}^d \times (0,\infty)} p_\infty \nabla \varphi . \nabla p_\infty = 0$$

which makes sense with the available regularity for p_∞ in Proposition 2.1.

The proof of Theorem 2.4 relies on a functional analysis argument which uses the L^∞ control from below for $\frac{\partial}{\partial t} n_\gamma \geq 0$ as given in Proposition 2.2.

2.6.1.3 From the Weak Formulation to the Free Boundary Statement

To begin with, notice that $\mathbb{1}_{\{\Omega(t)\}} = \mathbb{1}_{\{n_\infty(x,t)=1\}}$. Indeed, on the one hand, $\mathbb{1}_{\{\Omega(t)\}} \subset \mathbb{1}_{\{n_\infty(x,t)=1\}}$. On the other hand, when $p_\infty = 0$, then from (2.50), we conclude that $\frac{\partial}{\partial t} n_\infty = n_\infty G_M$, which means that we cannot have $n_\infty(x, t) = 1$ otherwise n_∞ would continue to grow thus contradicting the bound $n_\infty(x, t) \leq 1$.

Therefore, when $n_\infty(x, t)$ takes the values 0 or 1 only, then we have

$$n_\infty(x, t) = \mathbb{1}_{\{\Omega(t)\}}. \tag{2.53}$$

In this situation and assuming some smoothness for $\Omega(t)$, it is easy to derive the Hele-Shaw free boundary formulation mentioned in Sect. 2.5.1. This is written in details (and in more generality in the sense below) when $\Omega(t)$ is a ball in [77], then one can establish precisely the speed of the free boundary given by (2.37). The general study, including regularizing effects for the boundary $\partial\Omega(t)$ is performed in [66]. The difficulty is that different types of singularities may occur and new islands can be generated by places where $0 < n^0 < 1$, see also [61].

However, the weak formulation contains more than the free boundary statements (2.37), (2.38) which only holds true when initially $n^0 = \mathbb{1}_{\{\Omega(t=0)\}}$ so as to ensure (2.53). One can formally see this, because in the interior of $\Omega(t)$, we can write $\frac{\partial}{\partial t} n_\infty = 0$ and thus the weak formulation (2.50) gives immediately the elliptic equation (2.38). But, if there is a zone where $n^0 < 1$, then we still have $n_\infty(x, t) < 1$ for some time. In this space-time zone, we have $p_\infty = 0$ and (2.50) is reduced to the simple differential equation

$$\frac{\partial}{\partial t} n_\infty = n_\infty G_M.$$

A numerical simulation, illustrating this interpretation of competition between exponential growth and free boundary motion, is displayed in Fig. 2.3.

2.6.2 Single Cell Type with Nutrient, Incompressible Limit

The weak limit of the single cell system with nutrient (2.39)–(2.40) has also been established in [77]. The result is as follows

Theorem 2.5 *Let Φ, Ψ satisfy (2.32), and (n^0, c^0) satisfy the hypotheses (2.26) and (2.33). Then, after extraction of subsequences, the density n_γ, the nutrient c_γ and the pressure p_γ converge for all $T > 0$ strongly in $L^1(Q_T)$ as $m \to \infty$ to limits $n_\infty, c_\infty, p_\infty \in BV(Q_T)$ that satisfy $0 \leq n_\infty \leq 1, 0 \leq c_\infty \leq c_B, 0 \leq p_\infty \leq p_h$, and*

$$\begin{cases} \partial_t n_\infty = \Delta p_\infty + n_\infty \Phi(p_\infty, c_\infty), \ n_\infty(0) = n^0, \\ \partial_t c_\infty = \Delta c_\infty - n_\infty \Psi(p_\infty, c_\infty) \quad c_\infty(0) = c^0, \end{cases} \tag{2.54}$$

in a distributional sense, with the relation

$$p_\infty(1 - n_\infty) = 0.$$

However the complementary relation $p_\infty(\Delta p_\infty - \Phi(p_\infty, c_\infty))$ is not established by lack of an estimate à la Aronson-Bénilan. Therefore the direct link with the geometric problem is not established either.

2.6.3 Open Problems

We recap here some open problems mentioned in this chapter. They are on the mathematical side and do not address the many modeling challenges that rely on new experimental observations.

The theory of weak solutions can be carried out for the case with nutrients, see [77], but the complementary relation (2.51) is not established. That is because it relies on the Aronson-Bénilan estimate which is not available for the nutrient model (2.31), either in the modified form of the Aronson-Bénilan estimate proposed in [77]. Notice that a possible idea, following [19, 52] is to control Δp from below in a weaker space than L^∞. Namely for a quantity as $w = \Delta p + G$ or $w = \Delta p_\infty - \Phi$, a control of the negative part $\|w_-\|_{L^p_{loc}}$ is proved.

For multispecies models, the existence in a general framework with degeneracy is not established, see Sect. 2.4.3 for further comments. Also, the incompressible limit $\gamma \to \infty$ is also an open problem. The bounds provided in [52] are not enough to investigate this question. However the one dimensional case is under investigation [19] based upon arguments from [31].

Another question is about different mobilities which, see [56, 63], where the parabolic aspects of the equation for $n = n_1 + n_2$ do not apply. This is a long standing problem.

Acknowledgments The author has received partial funding from the European Research Council (ERC) under the European Union's Horizon 2020 research and innovation programme (grant agreement No 740623) and partial funding from the ANR blanche project Kibord ANR-13-BS01-0004 funded by the French Ministry of Research.

References

1. J. Adler, Chemotaxis in bacteria. Science **153**, 708–716 (1966)
2. A. Agosti, P.F. Antonietti, P. Ciarletta, M. Grasselli, M. Verani, A Cahn-Hilliard type equation with application to tumor growth dynamics. Math. Methods Appl. Sci. **40**(18), 7598–7626 (2017)
3. W. Alt, Biased random walk models for chemotaxis and related diffusion approximations. J. Math. Biol. **9**, 147–177 (1980)
4. H.W. Alt, E. DiBenedetto, Nonsteady flow of water and oil through inhomogeneous porous media. Ann. Scuola Norm. Sup. Pisa Cl. Sci. (4) **12**(3), 335–392 (1985)
5. H.W. Alt, S. Luckhaus, A. Visintin, On nonstationary flow through porous media. Ann. Mat. Pura Appl. (4) **136**, 303–316 (1984)
6. D.G. Aronson, P. Bénilan, Régularité des solutions de l'équation des milieux poreux dans $\mathbf{R}^N$. C. R. Acad. Sci. Paris Sér. A-B **288**(2), A103–A105 (1979)
7. C. Bardos, R. Santos, R. Sentis, Diffusion approximation and computation of the critical size. Trans. Amer. Math. Soc. **284**(2), 617–649 (1984)
8. N. Bellomo, M. Winkler, A degenerate chemotaxis system with flux limitation: maximally extended solutions and absence of gradient blow-up. Commun. Part. Diff. Equ. **42**, 436–473 (2017)
9. N. Bellomo, N.K. Li., P.K. Maini, On the foundations of cancer modelling: selected topics, speculations, and perspectives. Math. Models Methods Appl. Sci. **4**, 593–646 (2008)
10. H.C. Berg, *E. coli in Motion* (Springer, Berlin, 2004)
11. M. Bertsch, D. Hilhorst, H. Izuhara, M. Mimura, A nonlinear parabolic-hyperbolic system for contact inhibition of cell-growth. Diff. Eqs. Appl. **4**, 137–157 (2012)
12. P. Biler, L. Corrias, J. Dolbeault, Large mass self-similar solutions of the parabolic-parabolic Keller-Segel model of chemotaxis. J. Math. Biol. **63**(1), 1–32 (2011)
13. A. Blanchet, J. Dolbeault, B. Perthame, Two-dimensional Keller-Segel model: optimal critical mass and qualitative properties of the solutions. Electron. J. Diff. Equ. **2006**(44), 1–32 (2006)
14. A. Blanchet, J.A. Carrillo, P. Laurençot, Critical mass for a Patlak-Keller-Segel model with degenerate diffusion in higher dimensions. Calc. Var. Partial Differential Equations **35**(2), 133–168 (2009)
15. E. Bouin, V. Calvez, G. Nadin, Propagation in a kinetic reaction-transport equation: travelling waves and accelerating fronts. Arch. Ration. Mech. Anal. **217**(2), 571–617 (2015)
16. N. Bournaveas, V. Calvez, Critical mass phenomenon for a chemotaxis kinetic model with spherically symmetric initial data. Ann. Inst. H. Poincaré Anal. Non Linéaire **26**(5), 1871–1895 (2009)
17. N. Bournaveas, V. Calvez, S. Gutièrrez, B. Perthame, Comm. Partial Diff. Equ. **33**, 79–95 (2008)

18. M.P. Brenner, L.S. Levitov, E.O. Budrene, Physical mechanisms for chemotactic pattern formation by bacteria. Biophys J. **74**, 1677–1693 (1998)
19. F. Bubba, C. Pouchol, B. Perthame, M. Schmidtchen, Incompressible limit for a two species model of tissue growth in one space dimension. Arch. Ration. Mech. Anal. **236**(2), 735–766 (2020)
20. E.O. Budrene, H.C. Berg, Dynamics of formation of symmetrical patterns by chemotactic bacteria. Nature **376**, 49–53 (1995)
21. H.M. Byrne, M. Chaplain, Growth of necrotic tumors in the presence and absence of inhibitors. Math. Biosci. **135**(2),187–216 (1996)
22. H. Byrne, D. Drasdo, Individual-based and continuum models of growing cell populations: a comparison. J. Math. Biol. **58**, 657–687 (2009)
23. H. Byrne, L. Preziosi, Modelling solid tumour growth using the theory of mixtures. Math. Med. Biol. **20**, 341–66 (2004)
24. H.M. Byrne, J.R. King, D.L.S. McElwain, L. Preziosi, A two-phase model of solid tumor growth. Appl. Math. Lett. **16**, 567–573 (2003)
25. H. Byrne, L. Preziosi, Modelling solid tumour growth using the theory of mixtures. Math. Med. Biol. **20**(4), 341–366 (2003)
26. V. Calvez, Chemotactic waves of bacteria at the mesoscale. arXiv:1607.00429 (2016)
27. V. Calvez, J.A. Carrillo, Refined asymptotics for the subcritical Keller-Segel system and related functional inequalities. Proc. Amer. Math. Soc. **140**(10), 3515–3530 (2012)
28. V. Calvez, G. Raoul, C. Schmeiser, Confinement by biased velocity jumps: aggregation of Escherichia coli. Kinet. Relat. Models **8**(4), 651–666 (2015)
29. V. Calvez, B. Perthame, S. Yasuda, Traveling wave and aggregation in a flux-limited Keller-Segel model. Kinet. Relat. Models **11**(4), 891–909 (2018)
30. C. Cancès, T.O. Gallouët, L. Monsaingeon, Incompressible immiscible multiphase flows in porous media: a variational approach. Anal. PDE **10**(8), 1845–1876 (2017)
31. J.A. Carrillo, S. Fagioli, F. Santambrogio, M. Schmidtchen, Splitting schemes and segregation in reaction cross-diffusion systems. SIAM J. Math. Anal. **50**(5), 5695–5718 (2018)
32. C. Cercignani, R. Illner, M. Pulvirenti, The mathematical theory of dilute gases. In: *Applied Mathematical Sciences*, vol. 106 (Springer, New York, 1994), viii+347 pp.
33. F. Chalub, P.A. Markowich, B. Perthame, C. Schmeiser, Kinetic models for chemotaxis and their drift-diffusion limits. Monatsh. Math. **142**, 123–141 (2004)
34. C. Chatelain, T. Balois, P. Ciarletta, M. Ben Amar, Emergence of microstructural patterns in skin cancer: a phase separation analysis in a binary mixture. New J. Phys. **13**(11), 115013 (2011)
35. A. Chertock, A. Kurganov, X. Wang, Y. Wu, On a chemotaxis model with saturated chemotactic flux. Kinetic Related Models **5**, 51–95 (2012)
36. P. Ciarletta, L. Foret, M. Ben Amar, The radial growth phase of malignant melanoma: multiphase modelling, numerical simulations and linear stability analysis. J. R. Soc. Interface **8**(56), 345–368 (2011)
37. T. Colin, D. Bresch, E. Grenier, B. Ribba, O. Saut, Computational modeling of solid tumor growth: the avascular stage. SIAM J. Sci. Comput. **32**(4), 2321–2344 (2010)
38. T. Colin, A. Iollo, D. Lombardi, O. Saut, System identification in tumor growth modeling using semi-empirical eigenfunctions. Math. Models Methods Appl. Sci. **22**(6), 1250003 (30 pp) (2012)
39. F. Cornelis, O. Saut, P. Cumsille, D. Lombardi, A. Iollo, J. Palussière, T. Colin, In vivo mathematical modeling of tumor growth from imaging date: Soon to come in the future? Diagn. Interv. Imaging **94**(6), 593–600 (2013)
40. M.G. Crandall, M. Pierre, Regularizing effects for $u_t = \Delta \phi(u)$. Trans. Am. Math. Soc. **274**(1), 159–168 (1982)
41. Y. Dolak, C. Schmeiser, Kinetic models for chemotaxis: hydrodynamic limits and spatiotemporal mechanisms. J. Math. Biol. 51, 595–615 (2005)
42. R. Eftimie, Hyperbolic and kinetic models for self-organized biological aggregations and movement: a brief review. J. Math. Biol. **65**, 35–75 (2012)

43. R. Erban, H. Othmer, From individual to collective behaviour in bacterial chemotaxis. SIAM J. Appl. Math. **65**(2), 361–391 (2004)
44. R. Erban, H. Othmer, Taxis equations for amoeboid cells. J. Math. Biol. **54**, 847–885 (2007)
45. J. Escher, G. Simonett, Classical solutions for Hele-Shaw models with surface tension. Adv. Diff. Equ. **2**(4), 619–642 (1997)
46. A. Friedman, A hierarchy of cancer models and their mathematical challenges. Discrete Contin. Dynam. Systems Ser. B **4**(1), 147–159 (2004)
47. S. Frigeri, M. Grasselli, E. Rocca, On a diffuse interface model of tumor growth. European J. Appl. Math. **26**, 215–243 (2015)
48. H. Garcke, K.F. Lam, R. Nürnberg, E. Sitka, A multiphase Cahn-Hilliard-Darcy model for tumour growth with necrosis. Math. Models Methods Appl. Sci. **28**(3), 525–577 (2018)
49. R. Glassey, *The Cauchy Problem in Kinetic Theory* (SIAM, Philadelphia, 1996)
50. I. Golding, Y. Kozlovsky, I. Cohen, E. Ben Jacob, Studies of bacterial branching growth using reaction–diffusion models for colonial development. Physica A **260**, 510–554 (1998)
51. A. Goriely, The mathematics and mechanics of biological growth. In: *Interdisciplinary Applied Mathematics*, vol. 45 (Springer, New York, 2017)
52. P. Gwiazda, B. Perthame, A. Świerczewska-Gwiazda, A two species hyperbolic-parabolic model of tissue growth. Comm. Partial Diff. Equ. **44**(12), 1605–1618 (2019)
53. T. Hillen, H.G. Othmer, The diffusion limit of transport equations derived from velocity-jump processes. SIAM J. Appl. Math. **61**, 751–775 (2000)
54. T. Hillen, K.J. Painter. A user's guide to PDE models for chemotaxis. J. Math. Biol. **58**, 183–217 (2009)
55. T. Hillen, A. Swan, The diffusion limit of transport equations in biology. In: *Mathematical Models and Methods for Living Systems*. Lecture Notes in Mathematics, vol. 2167, pp. 73–129. Fond. CIME/CIME Found. Subser (Springer, Cham, 2016)
56. S. Hoehme, D. Drasdo, A cell-based simulation software for multi-cellular systems. Bioinformatics **26**(20), 2641–2642 (2010)
57. H.J. Hwang, K. Kang, A. Stevens, Global solutions of nonlinear transport equations for chemosensitive movement. SIAM. J. Math. Anal. **36**, 1177–1199 (2005)
58. F. James, N. Vauchelet, Chemotaxis: from kinetic equations to aggregate dynamics. Nonlinear Diff. Eq. Appl. **20**(1), 101–127 (2013)
59. T. Kawakami, Y. Sugiyama, Uniqueness theorem on weak solutions to the Keller-Segel system of degenerate and singular types. J. Diff. Equ. **260**(5), 4683–4716 (2016)
60. E.F. Keller, L.A. Segel, Traveling bands of chemotactic bacteria: A theoretical analysis. J. Theor. Biol. **30**, 235–248 (1971)
61. I. Kim, N. Požár, Porous medium equation to Hele-Shaw flow with general initial density. Trans. Amer. Math. Soc. **370**(2), 873–909 (2018)
62. J.-G. Liu, A. Lorz, A coupled chemotaxis-fluid model: global existence. Ann. Inst. H. Poincaré Anal. Non Linéaire **28**(5), 643–652 (2011)
63. T. Lorenzi, A. Lorz, B. Perthame, On interfaces between cell populations with different mobilities. Kinetic and Related Models **10**(1), 299–311 (2016)
64. J.S. Lowengrub, H.B. Frieboes, F. Jin, Y.-L. Chuang, X. Li, P. Macklin, S.M. Wise, V. Cristini, Nonlinear modelling of cancer: bridging the gap between cells and tumours. Nonlinearity **23**, R1–R91 (2010)
65. B. Mazzag, I. Zhulin, A. Mogilner, Model of bacterial band formation in aerotaxis. Biophys. J. **85**, 3558–3574 (2003)
66. A. Mellet, B. Perthame, F. Quiros, A Hele-Shaw problem for tumor growth. J. Funct. Anal. **273**, 306–3093 (2017)
67. N. Mittal, E.O. Budrene, M.P. Brenner, A. Van Oudenaarden, Motility of Escherichia coli cells in clusters formed by chemotactic aggregation. Proc. Natl. Acad. Sci. USA **100**, 13259–13263 (2003)
68. M. Mizoguchi, M. Winkler, Blow-up in the two-dimensional parabolic Keller-Segel system. Per. Commun.
69. J.D. Murray, *Mathematical Biology*, vol. 2, 2nd edn. (Springer, Berlin, 2002)

70. G. Nadin, B. Perthame, L. Ryzhik, Traveling waves for the Keller-Segel system with Fisher birth terms. Interface Free Bound **10**, 517–538 (2008)
71. H. Othmer, S. Dunbar, W. Alt, Models of dispersal in biological systems. J. Math. Biol. **26**, 263–298 (1988)
72. H.G. Othmer, T. Hillen, The diffusion limit of transport equations II: Chemotaxis equations. SIAM J. Appl. Math. **62**, 122–1250 (2002)
73. B. Perthame, Math. Tools for Kinetic Equations. Bull. Am. Math. Soc. **41**(2) (2004)
74. B. Perthame, Transport equations in biology. In: *Frontiers in Mathematics* (Birkhäuser Verlag, Basel, 2007), x+198 pp.
75. B. Perthame, N. Vauchelet, Z. Wang, The flux limited Keller-Segel system; properties and derivation from kinetic equations. Rev. Mat. Iberoam. **36**(2), 357–386 (2020)
76. B. Perthame, S. Yasuda, Stiff-response-induced instability for chemotactic bacteria and flux-limited Keller-Segel equation. Nonlinearity **31**, 4065–4089 (2018)
77. B. Perthame, F. Quiròs, J.-L. Vàzquez, The Hele-Shaw asymptotics for mechanical models of tumor growth. Arch. Ration. Mech. Anal. **212**(1), 93–127 (2014)
78. B. Perthame, M. Tang, N. Vauchelet, Traveling wave solution of the Hele–Shaw model of tumor growth with nutrient. Math. Models Methods Appl. Sci. **24**(13), 2601–2626 (2014)
79. B. Perthame, M. Tang, N. Vauchelet, Derivation of the bacterial run-and-tumble kinetic equation from a model with biochemical pathway. J. Math. Biol. **73**(5), 1161–1178 (2016)
80. A.B. Potapov, T. Hillen, Metastability in chemotaxis model. J. Dyn. Diff. Equat. **17**(2), 293–330 (2005)
81. L. Preziosi, A. Tosin, Multiphase modeling of tumor growth and extracellular matrix interaction: mathematical tools and applications. J. Math. Biol. **58**, 625–656 (2009)
82. L. Preziosi, G. Vitale, A multiphase model of tumor and tissue growth including cell adhesion and plastic reorganization. Math. Models Methods Appl. Sci. **21**(9), 1901–1932 (2011)
83. J. Ranft, M. Basan, J. Elgeti, J.-F. Joanny, J. Prost, F. Jülicher, Fluidization of tissues by cell division and apoptosis. Proc. Natl. Acad. Sci. U. S. A. **107**(49), 20863–20868 (2010)
84. B. Ribba, O. Saut, T. Colin, D. Bresch, E. Grenier, J. P. Boissel, A multiscale mathematical model of avascular tumor growth to investigate the therapeutic benefit of anti-invasive agents. J. Theoret. Biol. **243**(4), 532–541 (2006)
85. T. Roose, S. Chapman, P. Maini, Mathematical models of avascular tumour growth: a review. SIAM Rev. **49**(2), 179–208 (2007)
86. J. Saragosti, V. Calvez, N. Bournaveas, A. Buguin, P. Silberzan, B. Perthame, Mathematical description of bacterial traveling pulses. PLoS Comput. Biol. **6**(8), e1000890 (2010)
87. J. Saragosti, V. Calvez, N. Bournaveas, B. Perthame, A. Buguin, P. Silberzan, Directional persistence of chemotactic bacteria in a traveling concentration wave. Proc. Natl. Acad. Sci. **108**(39), 16235–16240 (2011)
88. J.A. Sherratt, M.A.J. Chaplain, A new mathematical model for avascular tumour growth. J. Math. Biol. **43**(4), 291–312 (2001)
89. G. Si, M. Tang, X. Yang, A pathway-based mean-field model for E. coli chemo-taxis: mathematical derivation and keller-segel limit. Multiscale Model Simul. **12**(2), 907–926 (2014)
90. M.J. Tindall, P.K. Maini, S.L. Porter, J.P. Armitage, Overview of mathematical approaches used to model bacterial chemotaxis II: bacterial populations. Bull Math Biol. **70**, 1570–1607 (2008)
91. J.-L. Vázquez, The porous medium equation. Mathematical theory. *Oxford Mathematical Monographs* (The Clarendon Press, Oxford University Press, Oxford, 2007). ISBN:978-0-19-856903-9
92. M. Winkler, Emergence of large population densities despite logistic growth restrictions in fully parabolic chemotaxis systems. Discrete Contin. Dyn. Syst. Ser. B **22**(7), 2777–2793 (2017)

Chapter 3
Segregated Algorithms for the Numerical Simulation of Cardiac Electromechanics in the Left Human Ventricle

L. Dede', A. Gerbi, and A. Quarteroni

Abstract We propose and numerically assess three segregated (partitioned) algorithms for the numerical solution of the coupled electromechanics problem for the left human ventricle. We split the coupled problem into its core mathematical models and we proceed to their numerical approximation. Space and time discretizations of the core problems are carried out by means of the Finite Element Method and Backward Differentiation Formulas, respectively. In our mathematical model, electrophysiology is represented by the monodomain equation while the Holzapfel-Ogden strain energy function is used for the passive characterization of tissue mechanics. A transmurally variable active strain model is used for the active deformation of the fibers of the myocardium to couple the electrophysiology and the mechanics in the framework of the active strain model. In this work, we focus on the numerical strategy to deal with the solution of the coupled model, which is based on novel segregated algorithms that we propose. These also allow using different time discretization schemes for the core submodels, thus leading to the formulation of staggered algorithms, a feature that we systematically exploit to increase the efficiency of the overall computational procedure. By means of numerical tests we show that these staggered algorithms feature (at least) first order of accuracy. We take advantage of the efficiency of the segregated schemes to solve, in a High Performance Computing framework, the cardiac electromechanics problem for the human left ventricle, for both idealized and subject-specific configurations.

L. Dede' (✉) · A. Gerbi
MOX–Modeling and Scientific Computing, Mathematics Department, Politecnico di Milano, Milano, Italy
e-mail: luca.dede@polimi.it

A. Quarteroni
MOX–Modeling and Scientific Computing, Mathematics Department, Politecnico di Milano, Milano, Italy

Institute of Mathematics, École Polytechnique Fédérale de Lausanne, Lausanne, Switzerland
e-mail: alfio.quarteroni@polimi.it

© The Editor(s) (if applicable) and The Author(s), under exclusive licence
to Springer Nature Switzerland AG 2020
A. DeSimone et al., *The Mathematics of Mechanobiology*,
Lecture Notes in Mathematics 2260, https://doi.org/10.1007/978-3-030-45197-4_3

81

3.1   Introduction

The heart performs two fundamental tasks: it pumps the deoxygenated blood to the lungs to get oxygen and release carbon dioxide, while it simultaneously pushes the oxygen rich blood into the arteries delivering it to tissues and organs [46], almost to every cell in the human body. Despite the apparent simplicity of these tasks, the heart function is however the result of the concerted action of several physical processes taking place at different spatial scales, i.e. at the cellular, tissue, and organ levels, other than time scales. In the mathematical modeling of the heart function all these processes have to be properly considered and, above all, integrated; we refer to this as an "integrated heart model" [62]. The electrophysiology, the active and the passive mechanics are referred to as "single core models", and are expressed by systems of Ordinary Differential Equations (ODEs) and Partial Differential Equations (PDEs). Although their individual behavior is nowadays quite established, further theoretical studies are necessary to better understand their interactions [12, 19, 26, 28, 61, 67]. Since, as noticed before, the processes under consideration feature different spatial and temporal scales, the grid for the numerical approximation of the individual core models must be properly chosen. Moreover, the discretized integrated problem can be formulated by either a *monolithic* approach, where the approximated equations are assembled in a single large system and simultaneously solved, or a *segregated* approach, where the approximated equations are solved sequentially.

We focus here on the electromechanics of the left ventricle (LV). For the investigation of this model and its numerical approximation we refer the interested reader to, e.g., [33, 53, 61, 62, 68, 77, 80, 82]. Segregated algorithms are investigated in [4, 12, 33, 44, 68, 82], where the electrophysiology and the mechanics problems are solved separately. In [20, 21, 37], the integrated problem is instead solved using a monolithic approach. In either case, suitable solvers (and preconditioners) must be employed for the efficient solution of the linear systems stemming from the discretization of the problems.

In this work we use the monodomain equation [13, 41, 59] together with the minimal Bueno-Orovio ionic model [8] for the description of the electrophysiology. For the passive mechanics, we use the state-of-the-art Holzapfel-Ogden model [39] together with the active strain approach [2, 3] for the active mechanics, the latter endowed with a newly proposed model for the transmurally heterogeneous thickening of the myocardium [6]. The mechanics is then coupled with the electrophysiology by means of a mathematical model describing the shortening of the myocardial fibers [28, 71], triggered by a change in the ionic concentrations in the cardiac cells, namely the intracellular calcium concentration. Regarding the numerical approximation of the integrated model, we use the Finite Element Method (FEM) for the space discretization while the time discretization is carried out by means of the Backward Differentiation Formulas (BDFs) [63]. We propose three novel segregated algorithms in combination with both implicit and semi-implicit schemes, the latter consisting in the partial evaluation of the nonlinear terms with an approximation of the unknowns of the same order of the BDF scheme [11, 29]. We compare the numerical results obtained by means of the segregated algorithms with those of the monolithic method proposed in [28] for physically meaningful benchmark problems. We develop our segregated algorithms in a way such that different timestep sizes for the electrophysiology and the mechanics can be used, thus leading to the so-called *staggered* algorithms. The use of different timestep sizes for the time discretization of the single core models is indeed made straightforward by the use of segregated approaches. Moreover, this is physically motivated as each core model features very different time scales: precisely, the electrophysiology requires a small timestep size while the mechanics yields stable and accurate results also for relatively coarse time discretizations. We show that the segregated schemes exhibit order of convergence with respect to the timestep size equal to one. Moreover, regarding the computational efficiency, we show that the segregated algorithms allow dramatic reductions of the computational costs with respect to the monolithic scheme. This is particularly true for a segregated algorithm in which the ionic, the monodomain, the mechanical activation, and the mechanics equations are fully decoupled and a timestep size ten times larger is used for the latter with respect to the former. Finally, we use the proposed algorithms for subject-specific large scale simulations for a full heartbeat and discuss the results thus obtained.

This chapter is organized as follows: in Sect. 3.2 we recall the mathematical models for the electrophysiology, the mechanics and the mechanical activation of the myocardium; in Sect. 3.3 we carry on the space and time discretizations of the single core models; in Sect. 3.4 we propose the segregated algorithms for the solution of the integrated problem; in Sect. 3.5 we report and discuss the numerical results obtained with the proposed methods; finally, we draw our Conclusions.

3.2   Mathematical Models

We recall, for each physical process, the underlying mathematical models in the form of ODEs and PDEs.

3.2.1   Ionic Model and Monodomain Equation

The systolic phase of the LV starts when the electric signal originated from the atrioventricular node is conveyed through the Purkinje fibers network and delivered to the myocardium [7, 14, 38, 55]. The signal triggers a complex interaction between the transmembrane potential v and different ionic species thus causing a quick depolarization and repolarization of the cells. This change of v is known as *action potential* [47]. The electric signal propagates faster along the myocardial fibers which, together with the sheets, characterizes the internal structure of the muscle [73] as depicted in Fig. 3.1.

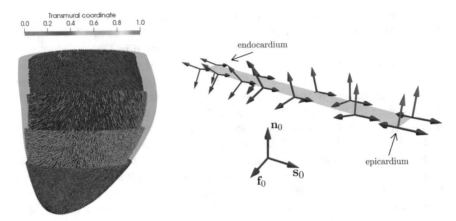

Fig. 3.1 Fibers in a subject-specific geometry (left) colored by a transmurally linear variable and a close up of a slice with fibers $\mathbf{f}_0$, sheets $\mathbf{s}_0$, and normals $\mathbf{n}_0$, together with their relative position (right)

We use a set of N_I ODEs in the ionic variables $\mathbf{w} = \{w_q\}_{q=1}^{N_I}$ to model the ionic species concentrations and currents through the cell membrane and the *monodomain equation*, a nonlinear diffusion-reaction parabolic equation derived from the bidomain equations under simplifying assumptions [13, 14], to model the tissue electrophysiology. This is compactly written as:

$$
\begin{cases}
\dfrac{\partial \mathbf{w}}{\partial t} = \boldsymbol{\alpha}(v)(\mathbf{w}^\infty(v) - \mathbf{w}) + \boldsymbol{\beta}(v)\mathbf{w} & \text{in } \Omega_0 \times (0, T), \\[2ex]
\chi\left(C_m \dfrac{\partial v}{\partial t} + I^{ion}(v, \mathbf{w})\right) = \nabla \cdot (J\mathbf{F}^{-1}\mathbf{D}_m\mathbf{F}^{-T}\nabla v) + I^{app}(t) & \text{in } \Omega_0 \times (0, T), \\[2ex]
(J\mathbf{F}^{-1}\mathbf{D}_m\mathbf{F}^{-T}\nabla v) \cdot \mathbf{N} = 0 & \text{on } \partial\Omega_0 \times (0, T), \\[2ex]
v = v_0, \quad \mathbf{w} = \mathbf{w}_0 & \text{in } \Omega_0 \times \{0\},
\end{cases}
\tag{3.1}
$$

Here, Ω_0 is the reference domain and $T > 0$ is the final time of the simulation; χ and $C_m \in \mathbb{R}^+$ are the ratio of membrane surface with respect to the volume and the membrane capacitance, respectively. The term $I^{ion}(v, \mathbf{w})$ represents the currents driven by the ions concentrations while I^{app} is an externally applied stimulus. The current geometry displacement (the displacement of the myocardium) $\mathbf{d} = \mathbf{X} - \mathbf{x}$ determines $\mathbf{F} = \mathbf{I} + \dfrac{\partial \mathbf{d}}{\partial \mathbf{X}}$ and $J = \det(\mathbf{F})$, where $\mathbf{X}$ and $\mathbf{x}$ are the space variables in the reference (Ω_0) and in the deformed (Ω) configurations, respectively. We neglect, for simplicity, stretch activated currents (SAC) [22, 42] and other bioelectrical effects of mechanical feedbacks [16–18]. To account for the anisotropic electrical conductance of the myocardium, the diffusion tensor reads $\mathbf{D}_m = \sigma_{iso}(\mathbf{I} - \mathbf{f}_0 \otimes \mathbf{f}_0) + \sigma_f \mathbf{f}_0 \otimes \mathbf{f}_0$, where $\mathbf{f}_0$ is the local fiber orientation (see Fig. 3.1) that varies transmurally, while the sheets direction $\mathbf{s}_0$ (which is oriented as the normal to the collagene sheets) is orthogonal to the LV walls. The $\mathbf{n}_0$ direction is orthogonal to both $\mathbf{f}_0$ and $\mathbf{s}_0$; σ_{iso} and σ_f are the fiber-transversal and the fiber-longitudinal conductances, respectively. Finally, the terms $\boldsymbol{\alpha}(v)$, $\boldsymbol{\beta}(v)$, and I^{ion} are prescribed according to the ionic model. Among the many models proposed in literature ([1, 47, 48, 52, 79]) we choose the Bueno-Orovio minimal model [8] for which $N_I = 3$. We assimilate the variable w_3 to the intracellular concentration $[Ca^{2+}]$.

3.2.2 Mechanical Activation

The concentration of calcium ions $[Ca^{2+}]$ drives complex dynamics in the sarcomeres [66], which lead to the cardiomyocites stretching. We use a phenomenological model for the local shortening of the fibers (denoted by γ_f) at the macroscopic level. The latter, firstly proposed in [71], and further developed in [28, 68], reads:

$$\begin{cases} \dfrac{\partial \gamma_f}{\partial t} - \dfrac{\varepsilon}{g(w_3)} \Delta \gamma_f = \dfrac{1}{g(w_3)} \Phi(w_3, \gamma_f, \mathbf{d}) & \text{in } \Omega_0 \times (0, T), \\[2ex] \nabla \gamma_f \cdot \mathbf{N} = 0 & \text{on } \partial\Omega_0 \times (0, T), \\[2ex] \gamma_f = 0 & \text{in } \Omega_0 \times \{0\}. \end{cases} \tag{3.2}$$

Here $g(w_3) = \widehat{\mu}_A w_3^2$, while $\Phi(w_3, \gamma_f, \mathbf{d})$ depends on the sarcomere force-length relationship [28, 34, 68]; we refer the reader to [28] for the expression of $\Phi(w_3, \gamma_f, \mathbf{d})$. Finally, $\widehat{\mu}_A, \varepsilon \in \mathbb{R}^+$ are tuning parameters for the subject-specific case under study.

3.2.3 Passive and Active Mechanics

We model the myocardium as a hyperelastic material [54]; then $\mathbf{P} = \mathbf{P}(\mathbf{d}) = \frac{\partial W}{\partial \mathbf{F}}$ is the first Piola–Kirchhoff stress tensor and W the strain energy density function. To model moderate volumetric changes (2–15%), we use the *nearly-incompressible* formulation [25] by multiplicatively decomposing the deformation gradient $\mathbf{F}$ into the isochoric $\overline{\mathbf{F}}$ and the volumetric $\mathbf{F}_v$ parts as $\mathbf{F} = \mathbf{F}_v \overline{\mathbf{F}}$, where $\mathbf{F}_v = J^{\frac{1}{3}} \mathbf{I}$. We use the Holzapfel-Ogden strain energy density function [39]—the state-of–the–art in passive myocardial tissue modeling—with a volumetric term weakly penalizing volumetric variations [74] $W(\mathbf{C}, J) = W_{el}(\mathbf{C}, J) + W_{vol}(J)$, where the expression of W_{el} is given in [28] with parameters determined in [39] and $W_{vol} = \frac{B}{2}(J - 1)\log(J)$; $B \in \mathbb{R}^+$ is the *bulk modulus*.

We use the *active strain* formulation [2, 3, 51, 70] to account for the active behavior of the myocardium. This approach corresponds to a decomposition of $\mathbf{F}$ in the form $\mathbf{F} = \mathbf{F}_E\mathbf{F}_A = J^{\frac{1}{3}}\overline{\mathbf{F}}_E\mathbf{F}_A$; $\overline{\mathbf{F}}_E$ is the isochoric component of the elastic (passive) part of the deformation, where $\mathbf{F}_E = \mathbf{F}_v\overline{\mathbf{F}}_E$, and $\mathbf{F}_A$ is the prescribed active deformation (active strain) tensor. We have that $\mathbf{P} = \det(\mathbf{F}_A)\mathbf{P}_E\mathbf{F}_A^{-T}$, where $\mathbf{P}_E = \frac{\partial W}{\partial \mathbf{F}_E}$. We refer to [28, 30, 31] for more details. We use the following orthotropic form for the tensor $\mathbf{F}_A$ [2, 5, 57, 68, 69]:

$$\mathbf{F}_A = \mathbf{I} + \gamma_f\mathbf{f}_0 \otimes \mathbf{f}_0 + \gamma_s\mathbf{s}_0 \otimes \mathbf{s}_0 + \gamma_n\mathbf{n}_0 \otimes \mathbf{n}_0,$$

where γ_n and γ_s are the local shortening (or elongation) of the tissue in the directions $\mathbf{s}_0$ and $\mathbf{n}_0$, respectively. Following [6, 28], we set γ_n as transmurally variable, $\gamma_n = k'(\lambda)\left(\frac{1}{\sqrt{1+\gamma_f}} - 1\right)$, where λ is a transmural coordinate, varying from λ_{endo} at the endocardium and λ_{epi} at the epicardium. The dependent variable γ_s is chosen to ensure $\det(\mathbf{F}_A) = 1$; hence, $\gamma_n = \gamma_n(\gamma_f)$ and $\gamma_s = \gamma_s(\gamma_f)$.

We finally use the stress tensor $\mathbf{P}(\mathbf{d}, \gamma_f)$ in the momentum equation:

$$
\begin{cases}
\rho\dfrac{\partial^2\mathbf{d}}{\partial t^2} - \nabla_0 \cdot \mathbf{P}(\mathbf{d}, \gamma_f) = \mathbf{0} & \text{in } \Omega_0 \times (0, T), \\[2ex]
(\mathbf{N} \otimes \mathbf{N})\left(K_\perp^\eta\mathbf{d} + C_\perp^\eta\dfrac{\partial\mathbf{d}}{\partial t}\right) & \\[2ex]
\qquad + (\mathbf{I} - \mathbf{N} \otimes \mathbf{N})\left(K_\parallel^\eta\mathbf{d} + C_\parallel^\eta\dfrac{\partial\mathbf{d}}{\partial t}\right) + \mathbf{P}(\mathbf{d}, \gamma_f)\,\mathbf{N} = \mathbf{0} & \text{on } \Gamma_0^\eta \times (0, T), \\[2ex]
\mathbf{P}(\mathbf{d}, \gamma_f)\,\mathbf{N} = p^{endo}(t)\mathbf{N} & \text{on } \Gamma_0^{endo} \times (0, T), \\[2ex]
\mathbf{d} = \mathbf{d}_0, \quad \dfrac{\partial\mathbf{d}}{\partial t} = \dot{\mathbf{d}}_0 & \text{in } \Omega_0 \times \{0\}.
\end{cases}
\tag{3.3}
$$

where ρ is the density of the myocardium. The boundary $\partial\Omega_0$ is partitioned in Γ_0^{endo}, Γ_0^{epi}, and Γ_0^{base}, representing the endocardium, the epicardium, and the ventricle base, respectively. For $\eta \in \{base, epi\}$, we consider generalized Robin conditions with parameters $K_\perp^\eta, K_\parallel^\eta, C_\perp^\eta, C_\parallel^\eta \in \mathbb{R}^+$, whereas the pressure $p^{endo}(t)$ (still prescribed at this stage) is set at the endocardium; the generalized Robin conditions are meant to represent the effect of the pericardium and surrounding tissues on the epicardium. Finally, $\mathbf{d}_0$ and $\dot{\mathbf{d}}_0$ denote initial conditions.

3.2.3.1 Prestress

If, at the initial time $t = 0$, $\overline{p}^{endo} = p^{endo}(0) > 0$, the blood pressure acts at the endocardial walls and thus the net force acting on the myocardium is not zero. The reference configuration Ω_0 is therefore unstressed at $t = 0$, which leads to unphysical deformations in problem (3.3). To overcome this issue, we use a *prestress* approach [40, 78] to determine the internal stresses of the myocardium such that the latter remains in mechanical equilibrium at $t = 0$. After observing that $\mathbf{P}(\mathbf{d}, \gamma_f) = \mathbf{P}(\mathbf{d}_0)$ at $t = 0$, we look for a vector $\widehat{\mathbf{d}}_0$ and a tensor $\mathbf{P}_0$ such that

$$
\begin{cases}
\nabla_0 \cdot \mathbf{P}(\widehat{\mathbf{d}}_0) = -\nabla_0 \cdot \mathbf{P}_0 & \text{in } \Omega_0, \\
(\mathbf{N} \otimes \mathbf{N}) K_\perp^\eta \widehat{\mathbf{d}}_0 + (\mathbf{I} - \mathbf{N} \otimes \mathbf{N}) K_\parallel^\eta \widehat{\mathbf{d}}_0 + \mathbf{P}(\widehat{\mathbf{d}}_0) \mathbf{N} = \mathbf{0} & \text{on } \Gamma_0^\eta, \\
\mathbf{P}(\widehat{\mathbf{d}}_0) \mathbf{N} = \overline{p}^{endo} \mathbf{N} & \text{on } \Gamma_0^{endo},
\end{cases}
\tag{3.4}
$$

with $\widehat{\mathbf{d}}_0 \approx \mathbf{d}_0$. We then use the stress tensor $\widetilde{\mathbf{P}}(\mathbf{d}, \gamma_f) = \mathbf{P}(\mathbf{d}, \gamma_f) + \mathbf{P}_0$ in place of $\mathbf{P}(\mathbf{d}, \gamma_f)$ in the first of Eq. (3.3), and set $\mathbf{d}_0 = \widehat{\mathbf{d}}_0$, $\dot{\mathbf{d}}_0 = \mathbf{0}$. Since the pair $(\widehat{\mathbf{d}}_0, \mathbf{P}_0)$ is a solution of Eq. (3.4), this allows the myocardium to remain in mechanical equilibrium at $t = 0$.

3.2.4   Cardiac Cycle

As we aim at modeling the LV electromechanics for a full heartbeat (typically about 0.8 s long), we need to account for the blood interaction with the LV along the different phases of the heartbeat (see Fig. 3.2). Hence, we solve different 0D model (ODEs) [26, 68, 82]. The phases are, in order:

1. Isovolumic contraction: the early stages of the LV contraction drive an increment of the endocardial pressure p^{endo} from the End Diastolic Pressure (EDP) p_{EDP}^{endo} (about 10 mmHg) to the one in the aorta p^{ao} (about 85 mmHg). We determine p^{endo} as the solution of

$$
\frac{dV^{endo}}{dt}(p^{endo}) = 0, \qquad t \in (0, T_1],
\tag{3.5}
$$

where $V^{endo}(0)$ is set to the initial LV volume. Thus, we require that the ventricular volume V^{endo} remains constant; $T_1 = T_1(p^{endo})$ is the earliest time occurrence at which $p^{endo} \geq p^{ao}$;
2. Ejection: the ventricular volume V^{endo} decreases due to the contraction of the LV forcing the blood to flow through the aortic valve. We use a two elements Windkessel 0D model [84] in the form:

$$
C \frac{dp^{endo}}{dt} = -\frac{p^{endo}}{R} - \frac{dV^{endo}}{dt}, \qquad t \in (T_1, T_2],
\tag{3.6}
$$

with $p^{endo}(T_1) = p^{ao}$ where C and R represent the capacitance and resistance of the electric circuit mimicking the blood flow in the aorta. This phase ends when

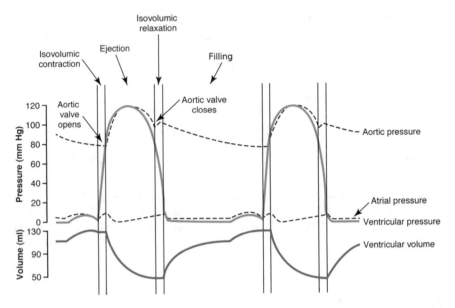

Fig. 3.2 Wiggers diagram [43] of the left heart depicting the aortic, ventricular, and atrial pressures as well as the ventricular volume, along the four phases of the cardiac cycle

p^{endo} becomes smaller than p^{ao}, thus closing the aortic valve. Since we do not model the aortic pressure over time, we set in Eq. (3.6) $T_2 = T_2(V^{endo})$ as soon as $\frac{dV^{endo}}{dt}(T_2) \geq 0$;

3. Isovolumic relaxation: the endocardial pressure p^{endo} decreases as a consequence of the LV early relaxation while V^{endo} remains constant and is treated similarly to the isovolumic contraction (Eq. (3.5)). We denote the end time of this phase as $T_3 = T_3(p^{endo})$, the occurrence at which $p^{endo} \leq p_{min}^{endo}$ (about 5 mmHg);

4. Filling: the pressure drop in the LV causes the opening of the mitral valve, which in turn causes an increment of V^{endo} due to the blood flowing into the LV, until both the pressure p^{endo} and the volume V^{endo} reach the EDP values. We model this phase by linearly increasing p^{endo} until it reaches the value p_{EDP}^{endo} at the time $\overline{T}_3 = 0.7\,\text{s}$, and we keep it constant from $\overline{T}_3$ to the final time $T = 0.8\,\text{s}$, that is:

$$\frac{dp^{endo}}{dt} = \varsigma, \qquad t \in (T_3, T], \tag{3.7}$$

with $\varsigma = \frac{p_{EDP}^{endo} - p^{endo}(T_3)}{\overline{T}_3 - T_3}$ if $t \in (T_3, \overline{T}_3]$ and $\varsigma = 0$ if $t \in (\overline{T}_3 T]$.

3.3 Space and Time Discretizations

We briefly discuss the numerical discretization of the single core models (3.1), (3.2), and (3.3) with respect to the space and the time independent variables.

3.3.1 Space Discretization

We use the Finite Element Method, FEM [63], for the space discretization of the PDEs of Sect. 3.2, thus obtaining a system of ODEs for each core model. We consider a mesh composed of pairwise disjoint tetrahedra $\mathcal{T}_h$ such that $\cup_{K \in \mathcal{T}_h} K = \Omega_0$, where h is the maximum size of the elements $K \in \mathcal{T}_h$. Then, we define the finite dimensional spaces $X_h^r = \{v \in C^0(\overline{\Omega}_0) : v|_K \in \mathbb{P}^r(K) \; \forall K \in \mathcal{T}_h\}$ and $\mathcal{X}_h^r = [X_h^r]^3$, where $\mathbb{P}^r(K)$ is the set of polynomials of degree smaller than or equal to r in the element K. By indicating with $\{\psi_j\}_{j=1}^{N_r^{\mathrm{dof}}}$ a basis for X_h^r, it holds $X_h^r = \mathrm{span}(\psi_1, \ldots, \psi_{N^{\mathrm{dof}}})$, $\mathcal{X}_h^r = \mathrm{span}(\{\boldsymbol{\psi}_1^k\}_{k=1}^3, \ldots, \{\boldsymbol{\psi}_{N_r^{\mathrm{dof}}}^k\}_{k=1}^3)$, and $\boldsymbol{\psi}_j^k = \psi_j \mathbf{e}_k$, where $\mathbf{e}_k$ is the k-th unit vector of $\mathbb{R}^3$. We then denote by $\{\mathbf{x}_j\}_{j=1}^{N^{\mathrm{dof}}}$ the set of the Degrees of Freedom (DoFs) associated to X_h^r and $\mathcal{X}_h^r$. The functions v_h, $\gamma_{f,h}$, and $\mathbf{d}_h$ are the FEM approximations of v, γ_f, and $\mathbf{d}$, respectively, and we denote by $\overline{\mathbf{v}}$, $\overline{\boldsymbol{\gamma}}_f$, and $\overline{\mathbf{d}}$ the vectors containing the nodal values of the primitive variables. Specifically, for the ionic variables $w_h^q(\mathbf{x}, t) = \sum_{j=1}^{N_r^{\mathrm{dof}}} w_j^q(t)\psi_j(\mathbf{x})$, where $w_j^q(t) \approx w_q(\mathbf{x}_j, t)$, from which $\overline{\mathbf{w}}^q(t) = \{w_j^q(t)\}_{j=1}^{N_r^{\mathrm{dof}}}$ and $\overline{\mathbf{w}}(t) = \{\overline{\mathbf{w}}^q(t)\}_{q=1}^{N_I}$. Similarly, for the transmembrane potential $v_h(\mathbf{x}, t) = \sum_{j=1}^{N_r^{\mathrm{dof}}} v_j(t)\psi_j(\mathbf{x})$, where $v_j(t) \approx v(\mathbf{x}_j, t)$ and $\overline{\mathbf{v}}(t) = \{v_j(t)\}_{j=1}^{N^{\mathrm{dof}}}$. For the active strain, $\gamma_{f,h}(\mathbf{x}, t) = \sum_{j=1}^{N_r^{\mathrm{dof}}} \gamma_{f,j}(t)\psi_j(\mathbf{x})$, where $\gamma_{f,j}(t) \approx \gamma_f(\mathbf{x}_j, t)$ and $\overline{\boldsymbol{\gamma}}_f(t) = \{\gamma_{f,j}(t)\}_{j=1}^{N^{\mathrm{dof}}}$. For the displacement, $\mathbf{d}_h(\mathbf{x}, t) = \sum_{j=1}^{N_r^{\mathrm{dof}}} \sum_{k=1}^3 d_j^k(t)\boldsymbol{\psi}_j^k(\mathbf{x})$, where $d_j^k(t) \approx \mathbf{d}(\mathbf{x}_j, t) \cdot \mathbf{e}_k$ and $\overline{\mathbf{d}}(t) = \{\{d_j^k(t)\}_{j=1}^{N_r^{\mathrm{dof}}}\}_{k=1}^3$.

We write the equations of the ionic model at each node $\mathbf{x}_j$, $j = 1, \ldots, N^{\mathrm{dof}}{}_r$. The semi-discrete formulation of the ionic model hence reads: given $\overline{\mathbf{v}}(t)$, find $\overline{\mathbf{w}}(t)$ such that

$$\begin{cases} \dot{\overline{\mathbf{w}}}(t) + \mathbb{U}(\overline{\mathbf{v}}(t))\overline{\mathbf{w}}(t) = \mathbf{Q}(\overline{\mathbf{v}}(t)), & t \in (0, T], \\ \overline{\mathbf{w}}(0) = \overline{\mathbf{w}}_0, \end{cases} \tag{3.8}$$

where $\mathbb{U}_{ii}(\overline{\mathbf{v}}) = \alpha_q(v_j) - \beta_q(v_j)$ and $\mathbf{Q}_i(\overline{\mathbf{v}}) = \alpha_q(v_j)w_q^\infty(v_j)$, with $i = q N_r^{\mathrm{dof}} + j$, for $q = 1, \ldots, N_I$, $j = 1, \ldots, N_r^{\mathrm{dof}}$.

For the monodomain equation we obtain instead the following semi-discrete problem: given $\overline{\mathbf{w}}(t)$ and $\overline{\mathbf{d}}(t)$, find $\overline{\mathbf{v}}(t)$ such that

$$\begin{cases} \mathbb{M}\dot{\overline{\mathbf{v}}}(t) + \mathbb{K}(\overline{\mathbf{d}}(t))\overline{\mathbf{v}}(t) + \mathbf{I}^{ion}(\overline{\mathbf{v}}(t), \overline{\mathbf{w}}(t)) = \mathbb{M}\mathbf{I}^{app}(t), \quad t \in (0, T], \\ \overline{\mathbf{v}}(0) = \overline{\mathbf{v}}_0, \end{cases} \tag{3.9}$$

where $\mathbb{M}_{ij} = \int_{\Omega_0} \psi_j \psi_i \, d\Omega_0$, $\mathbb{K}_{ij}(\overline{\mathbf{d}}_h) = \int_{\Omega_0} (J_h \mathbf{F}_h^{-1} \mathbf{D}_m \mathbf{F}_h^{-T} \nabla \psi_j) \cdot \nabla \psi_i \, d\Omega_0$, $\mathbf{I}_i^{ion}(\overline{\mathbf{v}}, \overline{\mathbf{w}}) = \int_{\Omega_0} I^{ion}\left(v_h, w_h^1, \ldots, w_h^{N_I}\right) \psi_i \, d\Omega_0$, and $\mathbf{I}_i^{app}(t) = I^{app}(\mathbf{x}_i, t)$ for $i, j = 1, \ldots, N_r^{\text{dof}}$ and $\overline{\mathbf{v}}_0 = \{v_0(\mathbf{x}_j)\}_{j=1}^{N_r^{\text{dof}}}$; $\mathbf{F}_h = \frac{\partial \mathbf{d}_h}{\partial \mathbf{X}}$ and $J_h = \det(\mathbf{F}_h)$. To overcome numerical instabilities [9], we use however a lumped mass matrix $\mathbb{M}^L$ in place of $\mathbb{M}$ [60].

The semi-discrete formulation of the mechanical activation problem (3.2) reads: given $\overline{\mathbf{w}}(t)$ and $\overline{\mathbf{d}}(t)$, find $\overline{\boldsymbol{\gamma}}_f(t)$ such that

$$\begin{cases} \mathbb{M}\dot{\overline{\boldsymbol{\gamma}}}_f(t) + \varepsilon \mathbb{K}(\overline{\mathbf{w}}(t))\overline{\boldsymbol{\gamma}}_f(t) + \boldsymbol{\Phi}(\overline{\mathbf{w}}(t), \overline{\boldsymbol{\gamma}}_f(t), \overline{\mathbf{d}}(t)) = \mathbf{0} \quad t \in (0, T], \\ \overline{\boldsymbol{\gamma}}_f(0) = \mathbf{0}, \end{cases} \tag{3.10}$$

where $\mathbb{K}_{ij}(\overline{\mathbf{w}}) = \int_{\Omega_0} \frac{\varepsilon}{g(w_h^3)} \nabla \psi_j \cdot \nabla \psi_i \, d\Omega_0$ and $\boldsymbol{\Phi}_i(\overline{\mathbf{w}}, \overline{\boldsymbol{\gamma}}_f, \overline{\mathbf{d}}) = -\int_{\Omega_0} \frac{1}{g(w_h^3)} \Phi \left(w_h^3, \gamma_{f,h}, \mathbf{d}_h\right) \psi_i \, d\Omega_0$, for $i, j = 1, \ldots, N_r^{\text{dof}}$.

The semi-discrete formulation of the mechanics problem (3.3) reads: given $\overline{\boldsymbol{\gamma}}_f(t)$, find $\overline{\mathbf{d}}(t)$ such that

$$\begin{cases} \rho_s \mathbb{M}_3 \ddot{\overline{\mathbf{d}}}(t) + \mathbb{F}\dot{\overline{\mathbf{d}}}(t) + \mathbb{G}\overline{\mathbf{d}}(t) + \mathbf{S}(\overline{\mathbf{d}}(t), \overline{\boldsymbol{\gamma}}_f(t)) = \mathbf{p}^{endo}(t) - \mathbf{S}_0, \quad t \in (0, T], \\ \overline{\mathbf{d}}(0) = \overline{\mathbf{d}}_0, \quad \dot{\overline{\mathbf{d}}}(0) = \mathbf{0}, \end{cases} \tag{3.11}$$

where $\mathbf{d}_0 = \left\{\{\mathbf{d}_0(\mathbf{x}_j) \cdot \mathbf{e}_k\}_{j=1}^{N_r^{\text{dof}}}\right\}_{k=1}^3$, $\mathbb{M}_3 = \text{diag}\{\mathbb{M}, \mathbb{M}, \mathbb{M}\}$, $\mathbf{p}_i^{endo} = \int_{\Gamma_0^{endo}} p^{endo} \mathbf{N} \cdot \boldsymbol{\psi}_i$, $\mathbf{S}_{0,i} = \int_{\Omega_0} \mathbf{P}_0 : \nabla_0 \boldsymbol{\psi}_i^k \, d\Omega_0$, $\mathbb{F}_{ij}^k = \sum_{\eta \in \{epi, base\}} \int_{\Gamma_0^\eta} \left(C_\perp^\eta (\mathbf{N} \otimes \mathbf{N}) + C_\parallel^\eta (\mathbf{I} - \mathbf{N} \otimes \mathbf{N})\right) \boldsymbol{\psi}_j^k \cdot \boldsymbol{\psi}_i^k \, d\Gamma_0$, $\mathbb{F} = \text{diag}\{\mathbb{F}^k\}_{k=1}^3$, $\mathbb{G}_{ij}^k = \sum_{\eta \in \{epi, base\}} \int_{\Gamma_0^\eta} \left(K_\perp^\eta (\mathbf{N} \otimes \mathbf{N}) + K_\parallel^\eta (\mathbf{I} - \mathbf{N} \otimes \mathbf{N})\right) \boldsymbol{\psi}_j^k \cdot \boldsymbol{\psi}_i^k \, d\Gamma_0$, $\mathbb{G} = \text{diag}\{\mathbb{G}^k\}_{k=1}^3$, $\mathbf{S}_i^k(\overline{\mathbf{d}}, \overline{\boldsymbol{\gamma}}_f) = \int_{\Omega_0} \mathbf{P}(\mathbf{d}_h, \gamma_{f,h}) : \nabla_0 \boldsymbol{\psi}_i^k \, d\Omega_0$, $\mathbf{S}(\overline{\mathbf{d}}, \overline{\boldsymbol{\gamma}}_f) = \text{diag}\left\{\mathbf{S}^k(\overline{\mathbf{d}}, \overline{\boldsymbol{\gamma}}_f)\right\}_{k=1}^3$ for $i, j = 1, \ldots, N_r^{\text{dof}}$. The discretized prestress problem (3.4) can be similarly written as: find $(\widehat{\overline{\mathbf{d}}}_0, \overline{\mathbf{P}}_0)$ such that

$$\mathbb{G}\widehat{\overline{\mathbf{d}}}_0 + \mathbf{S}(\widehat{\overline{\mathbf{d}}}_{0,h}) = \overline{\mathbf{p}}_{EDP}^{endo} - \mathbf{S}_0. \tag{3.12}$$

3.3.2 Time Discretization

We now carry on the time discretization for each of the semi-discrete single core problems of Sect. 3.3.1. For the electrophysiology several approaches have been proposed and used to solve the monodomain equation, in combination with the ionic model: explicit schemes [36, 68], implicit schemes [49, 59, 85], and implicit-explicit (IMEX) schemes [15, 75]. Regarding the mechanics, we consider an implicit scheme, while we consider both implicit and semi-implicit schemes for the electrophysiology. We symbolically rewrite the semi-discrete problems of Sect. 3.2 in the general form:

$$
\begin{cases}
M_i \mathbf{z}_i(t) + \mathbf{T}_i(\mathbf{z}(t)) = \mathbf{H}_i(t) & t \in (0, T], \quad i = 1, \ldots, 4, \\
\mathbf{z}_i(0) = \mathbf{z}_{i,0}, & i = 1, \ldots, 4, \\
\dot{\mathbf{z}}_4(0) = \mathbf{0},
\end{cases}
\tag{3.13}
$$

where $\mathbf{z}_1 = \overline{\mathbf{w}}$, $\mathbf{z}_2 = \overline{\mathbf{v}}$, $\mathbf{z}_3 = \overline{\boldsymbol{\gamma}}_f$, $\mathbf{z}_4 = \overline{\mathbf{d}}$, and $M_1 = \mathbb{I}\frac{d}{dt}$, $M_2 = M_3 = \mathbb{M}\frac{d}{dt}$, $M_4 = \rho_s \mathbb{M}_3 \frac{d^2}{dt^2}$. The nonlinear vector-valued functions $\mathbf{T}_i$ and $\mathbf{H}_i$ are specific of the corresponding core model. In order to obtain a fully discretized formulation using the BDF scheme, we exploit the following approximation of the time derivatives:

$$
\frac{d}{dt}\mathbf{z}_i(t^{n+1}) \approx \frac{1}{\Delta t}\left(\vartheta_0^I \mathbf{z}_i^{n+1} - \mathbf{z}_i^I\right), \qquad \mathbf{z}_i^I = \sum_{k=1}^{\sigma} \vartheta_k' \mathbf{z}_i^{n-k+1}, \qquad i = 1, \ldots, 4
$$

$$
\frac{d^2}{dt^2}\mathbf{z}_4(t^{n+1}) \approx \frac{1}{(\Delta t)^2}\left(\vartheta_0^{II}\mathbf{z}_4^{n+1} - \mathbf{z}_4^{II}\right), \qquad \mathbf{z}_4^{II} = \sum_{k=1}^{\sigma+1} \vartheta_k^{II}\mathbf{z}_4^{n-k+1},
$$

$$
\tag{3.14}
$$

where $\Delta t = \frac{T}{N_T}$ is the timestep size, N_T being the number of subintervals, while the parameters ϑ_k', ϑ_k'', $k = 0, \ldots, \sigma$ depend on the order σ of the BDF scheme.

In the implicit case, we obtain the following nonlinear systems:

$$
\mathbf{A}_i(\mathbf{z}^{n+1}) = \mathbf{b}_i^{n+1}, \qquad i = 1, \ldots, 4, \quad n = \sigma, \ldots, N_T - 1, \tag{3.15}
$$

with $\mathbf{z}^n$ assigned for $n = 0, \ldots, \sigma$. In the semi-implicit case, on the other hand, we extrapolate the variables in the nonlinear terms $\mathbf{A}_i(\mathbf{z}^{n+1})$ by means of the Newton-Gregory backward polynomials [11]—as done, e.g., for the Navier–Stokes equations in [29]—thus yielding a linear system at each timestep. The extrapolated variables are evaluated by means of an approximation of the same order σ of the BDF scheme as $\mathbf{z}_i(t^{n+1}) \approx \mathbf{z}_i^* = \sum_{k=1}^{\sigma} \beta_k \mathbf{z}_i^{n-k+1}$. We thus approximate the nonlinear terms as $\mathbf{A}_i(\mathbf{z}^{n+1}) \approx \sum_{j=1}^{4} \mathbb{A}_{i,j}(\mathbf{z}^*)\mathbf{z}_j^{n+1} + \tilde{\mathbf{A}}_i(\mathbf{z}^*)$, with notation being understood. By recalling Eq. (3.15), we hence obtain a system in the form:

$$\sum_{j=1}^{4} \mathbb{A}_{i,j}(\mathbf{z}^*)\mathbf{z}_j^{n+1} = \mathbf{b}_i^{n+1} \qquad n = \sigma, \dots, N_T - 1, \tag{3.16}$$

with $\mathbf{z}^n$ assigned for $n = 0, \dots, \sigma$ and $\mathbf{b}_i^{n+1} = \mathbf{h}_i^{n+1} - \widetilde{\mathbf{A}}_i(\mathbf{z}^*)$.

3.3.2.1 Discretization of the 0D Fluid Model

We evaluate the volume $V^{endo}(t)$ at time t^n by exploiting the formula reported in [68]. For the discretization of the 0D fluid models of Sect. 3.2.4 in terms of p^{endo}, we consider the following approaches tailored on the phase of the heartbeat (we drop the "*endo*" superscript for simplicity). At each n:

1. Isovolumic contraction: we use an inexact Newton method to solve Eq. (3.5) by iteratively updating the pressure as $p_{k+1}^{n+1} = p_k^{n+1} - \frac{\Delta t}{\zeta}(V_k^{n+1} - V^n)$, for $k = 0, 1, \dots$ with $p_0^{n+1} = p^n$ and $V_0^{n+1} = V^n$. By dimensional arguments, we approximate $\frac{\partial V}{\partial p}(p_k^{n+1})$ as $-\frac{\Delta t}{\zeta} \left[\frac{mm^4 s^2}{g} \right]$ in the Newton iterate. At each iteration, p_{k+1}^{n+1} is used to solve the electromechanics problem thus obtaining V_{k+1}^{n+1}; the procedure is repeated until the condition $\frac{|V_{k+1}^{n+1} - V^n|}{\Delta t} < \varepsilon$ is satisfied. The parameter $\zeta < 0$ has to be "sufficiently" small in order for the fixed point algorithm to converge;

2. Ejection: the two elements Windkessel model (3.6) is solved in the pressure variable with a BDF scheme of order $\sigma = 1$:

$$C \frac{p^{n+1} - p^n}{\Delta t} = -\frac{p^{n+1}}{R} - \frac{V^n - V^{n-1}}{\Delta t}; \tag{3.17}$$

3. Isovolumic relaxation: we proceed as in 1);
4. Filling: the pressure is simply updated as $p^{n+1} = p^n + \Delta t \, \varsigma$.

3.4 Numerical Coupling: Segregated Strategies

We first recall the monolithic strategy that we introduced in [28], then we propose three new segregated strategies for the solution of the electromechanics problem.

3.4.1 Fully Monolithic Strategy ($\mathcal{I}_I\mathcal{E}_I\mathcal{A}_I\mathcal{M}_I$)

We use the implicit scheme (3.15) for the time discretization of each core model and we assemble the integrated problem in a monolithic fashion, thus considering a "strong" coupling among the fully discretized core models; see [28]. This amounts to solve, for $n = \sigma, \ldots, N_T - 1$, the following system of size $8 \times N_r^{\text{dof}}$:

$$(\mathcal{I}_I\mathcal{E}_I\mathcal{A}_I\mathcal{M}_I) : \begin{cases} \left(\dfrac{\vartheta_0^{\text{I}}}{\Delta t} + \mathbb{U}(\overline{\mathbf{v}}^{n+1})\right)\overline{\mathbf{w}}^{n+1} - \mathbf{Q}(\overline{\mathbf{v}}^{n+1}) = \dfrac{1}{\Delta t}\overline{\mathbf{w}}^{\text{I}}, \\[4mm] \left(\dfrac{\vartheta_0^{\text{I}}}{\Delta t}\mathbf{M} + \mathbb{K}(\overline{\mathbf{d}}^{n+1})\right)\overline{\mathbf{v}}^{n+1} + \mathbf{I}^{ion}(\overline{\mathbf{v}}^{n+1}, \overline{\mathbf{w}}^{n+1}) = \dfrac{1}{\Delta t}\mathbf{M}\overline{\mathbf{v}}^{\text{I}} + \mathbf{M}\mathbf{I}^{app}(t^{n+1}), \\[4mm] \left(\dfrac{\vartheta_0^{\text{I}}}{\Delta t}\mathbf{M} + \varepsilon\mathbb{K}(\overline{\mathbf{w}}^{n+1})\right)\overline{\boldsymbol{\gamma}}_f^{n+1} + \boldsymbol{\Phi}(\overline{\mathbf{w}}^{n+1}, \overline{\boldsymbol{\gamma}}_f^{n+1}, \overline{\mathbf{d}}^{n+1}) = \dfrac{1}{\Delta t}\mathbf{M}\overline{\boldsymbol{\gamma}}_f^{\text{I}}, \\[4mm] \left(\rho_s\dfrac{\vartheta_0^{\text{II}}}{(\Delta t)^2}\mathbb{M}_3 + \dfrac{\vartheta_0^{\text{I}}}{\Delta t}\mathbf{F} + \mathbb{G}\right)\overline{\mathbf{d}}^{n+1} + \mathbf{S}(\overline{\mathbf{d}}^{n+1}, \overline{\boldsymbol{\gamma}}_f^{n+1}) \\[4mm] \qquad = \rho_s\dfrac{1}{(\Delta t)^2}\mathbb{M}_3\overline{\mathbf{d}}^{\text{II}} + \dfrac{1}{\Delta t}\mathbf{F}\overline{\mathbf{d}}^{\text{I}} + \mathbf{p}^{endo}(t^{n+1}) - \mathbf{S}_0, \end{cases}$$

$$(3.18)$$

which we indicate as $(\mathcal{I}_I\mathcal{E}_I\mathcal{A}_I\mathcal{M}_I)$, where the subscript I stands for the implicit solver, and compactly rewrite in algebraic form as

$$\mathbf{A}_{EM}^{n+1}(\mathbf{z}^{n+1}) = \mathbf{b}_{EM}^{n+1}, \tag{3.19}$$

with notation being understood. We then apply, at each timestep, the Newton method [63] to approximate the solution of the nonlinear problem (3.19) by iteratively solving the linear system

$$\mathbb{J}_{EM,k}^{n+1}\delta\mathbf{z}_{k+1}^{n+1} = -\mathbf{r}_k^{n+1} \quad \text{with} \quad \mathbf{z}_{k+1}^{n+1} = \mathbf{z}_k^{n+1} + \delta\mathbf{z}_{k+1}^{n+1}, \tag{3.20}$$

for $k = 0, \ldots$, until $\|\mathbf{r}_k^{n+1}\|_{L^2} < \varepsilon_{tol}^N$, where ε_{tol}^N is a given tolerance. $\mathbb{J}_{EM,k}^{n+1}$ is the Jacobian matrix of (3.19), evaluated in $\mathbf{z}_k^{n+1}$, and is endowed with the following block structure:

$$\mathbb{J}_{EM} = \begin{bmatrix} \boxed{\begin{matrix} J_{11} & J_{12} \\ J_{21} & J_{22} \end{matrix}} & 0 & 0 \\[2mm] & 0 & J_{24} \\[2mm] J_{31} & 0 & \boxed{\begin{matrix} J_{33} & J_{34} \\ J_{43} & J_{44} \end{matrix}} \\ 0 & 0 & \end{bmatrix}$$

$$(3.21)$$

while the residual is defined as $\mathbf{r}_k^{n+1} = \mathbf{b}_{EM}^{n+1} - \mathbf{A}_{EM}^{n+1}(\mathbf{z}_k^{n+1})$. In (3.21) we highlight the diagonal blocks corresponding to the electrophysiology, the mechanical activation, and the mechanics, respectively. We use the preconditioned GMRES method [72] to solve problem (3.20). We exploit a lower block triangular Gauss–Seidel right preconditioner $\mathcal{P}_{EM}$ introduced in [28], a generalization of the FaCSI preconditioner of [23, 24, 27]. $\mathcal{P}_{EM}$ is obtained by dropping the upper triangular blocks of matrix $\mathbb{J}_{EM}$, namely J_{12}, J_{24}, and J_{34}, and then by substituting the diagonal blocks with black-box Algebraic Multigrid (AMG) and Additive Schwarz preconditioners. With this strategy, we are able to exploit the information of the core problems at the block level, that is we use a preconditioner that exploits the "physics" of the coupled problem.

While $(\mathcal{I}_I \mathcal{E}_I \mathcal{A}_I \mathcal{M}_I)$ is "numerically" stable and convergent as long as the initial guess $\mathbf{z}_0^{n+1}$ in (3.20) is, at each time, "sufficiently" close to the solution, it also requires to use the same timestep for the time discretization of each core model. Hence, even if the electrophysiology and the mechanics feature very different time scales, the former dictates our choice for the timestep of the fully monolithic problem.

3.4.2 Partially Segregated Strategy $(\mathcal{I}_I \mathcal{E}_I \mathcal{A}_I)$–$(\mathcal{M}_I)$

We break the strong coupling between the electrophysiology and the mechanical activation $(\mathcal{I}_I \mathcal{E}_I \mathcal{A}_I)$ and the tissue mechanics $(\mathcal{M}_I)$. We hence evaluate the terms $\mathbb{K}(\overline{\mathbf{d}}^{n+1})$ and $\mathbf{\Phi}(\overline{\mathbf{w}}^{n+1}, \overline{\mathbf{\gamma}}_f^{n+1}, \overline{\mathbf{d}}^{n+1})$ of Eq. (3.18) in the extrapolated variable $\overline{\mathbf{d}}^*$ instead of $\overline{\mathbf{d}}^{n+1}$, thus obtaining two separated problems which are solved in a segregated fashion. This strategy is equivalent to the application of a (first order) Godunov splitting scheme [32] to the monolithic problem. We notice that the $(\mathcal{I}_I \mathcal{E}_I \mathcal{A}_I)$ problem is still fully coupled, while it is decoupled from the $(\mathcal{M}_I)$ block, hence the denomination $(\mathcal{I}_I \mathcal{E}_I \mathcal{A}_I)$–$(\mathcal{M}_I)$.

This approach allows to use a smaller timestep for the $(\mathcal{I}_I \mathcal{E}_I \mathcal{A}_I)$ problem, which we denote by τ, with respect to the one used for the mechanics $(\mathcal{M}_I)$: we set in particular

$$\tau = \frac{\Delta t}{N_{sub}},$$

where $N_{sub} \in \mathbb{N}$ is the number of intermediate substeps; τ is the timestep size of $(\mathcal{I}_I \mathcal{E}_I \mathcal{A}_I)$ and Δt that of $(\mathcal{M}_I)$. This implies that $\tau \leq \Delta t$ and $t^{n+\frac{m}{N_{sub}}} = t^n + m\tau$ for $m = 1, \ldots, N_{sub}$. N_{sub} can also be regarded as the ratio of the timestep lengths used for the mechanics and for the electrophysiology and activation. The overall time advancement is represented in Fig. 3.3. Another clear advantage of this approach is that, in the isovolumic phases, only the mechanics problem needs to be solved, contrarily to the fully monolithic one where Eq. (3.18) has to be solved at each

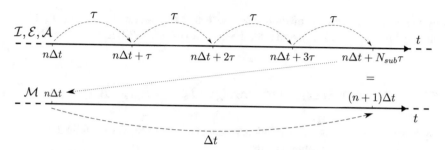

Fig. 3.3 Graphical representation of the time advancement for $(\mathcal{I}_I\mathcal{E}_I\mathcal{A}_I)$–$(\mathcal{M}_I)$ and $(\mathcal{I}_{SI}\mathcal{E}_{SI}\mathcal{A}_{SI})$–$(\mathcal{M}_I)$

subiteration. Problem $(\mathcal{I}_I\mathcal{E}_I\mathcal{A}_I)$ from t^n to t^{n+1} reads:

$$(\mathcal{I}_I\mathcal{E}_I\mathcal{A}_I): \begin{cases} \left(\dfrac{\vartheta_0^I}{\tau} + \mathbb{U}(\overline{\mathbf{v}}^{n+\frac{m}{N_{sub}}})\right)\overline{\mathbf{w}}^{n+\frac{m}{N_{sub}}} - \mathbb{Q}(\overline{\mathbf{v}}^{n+\frac{m}{N_{sub}}}) = \dfrac{1}{\tau}\overline{\mathbf{w}}^I, \\[2ex] \left(\dfrac{\vartheta_0^I}{\tau}\mathbb{M} + \mathbb{K}(\overline{\mathbf{d}}^*)\right)\overline{\mathbf{v}}^{n+\frac{m}{N_{sub}}} + \mathbf{I}^{ion}(\overline{\mathbf{v}}^{n+\frac{m}{N_{sub}}}, \overline{\mathbf{w}}^{n+\frac{m}{N_{sub}}}) = \dfrac{1}{\tau}\mathbb{M}\overline{\mathbf{v}}^I = \mathbb{M}\mathbf{I}^{app}(t^{n+\frac{m}{N_{sub}}}), \\[2ex] \left(\dfrac{\vartheta_0^I}{\tau}\mathbb{M} + \varepsilon\mathbb{K}(\overline{\mathbf{w}}^{n+\frac{m}{N_{sub}}})\right)\overline{\boldsymbol{\gamma}}_f^{n+\frac{m}{N_{sub}}} + \boldsymbol{\Phi}(\overline{\mathbf{w}}^{n+\frac{m}{N_{sub}}}, \overline{\boldsymbol{\gamma}}_f^{n+\frac{m}{N_{sub}}}, \overline{\mathbf{d}}^*) = \dfrac{1}{\tau}\mathbb{M}\overline{\boldsymbol{\gamma}}_f^I, \end{cases}$$

$$(3.22)$$

for $m = 1, \ldots, N_{sub}$, where the terms $\overline{\mathbf{w}}^I$, $\overline{\mathbf{v}}^I$, and $\overline{\boldsymbol{\gamma}}_f^I$ (defined in Eq. (3.14)) are evaluated by using the variables at times $t^n, t^n - \tau, \ldots, t^n - (\sigma - 1)\tau$. As in the case of the implicit electromechanics, we use the Newton method to solve problem (3.22), and the block structure of the correspondent Jacobian matrix $\mathbb{J}_{EA}$ is:

$$\mathbb{J}_{EA} = \begin{bmatrix} \begin{array}{cc|c} J_{11} & J_{12} & 0 \\ J_{21} & J_{22} & 0 \\ \hline J_{31} & 0 & \boxed{J_{33.}} \end{array} \end{bmatrix}$$

$$(3.23)$$

We exploit the same preconditioning technique that was outlined in Sect. 3.4.1 for the $(\mathcal{I}_I\mathcal{E}_I\mathcal{A}_I\mathcal{M}_I)$ strategy, but we restrict it to the block $(\mathcal{I}_I\mathcal{E}_I\mathcal{A}_I)$. After solving Eq. (3.22) for N_{sub} steps, we solve at t^{n+1} the implicit mechanics problem $(\mathcal{M}_I)$:

$$(\mathcal{M}_I): \begin{array}{l} \left(\rho_s\dfrac{\vartheta_0^{II}}{(\Delta t)^2}\mathbb{M}_3 + \dfrac{\vartheta_0'}{\Delta t}\mathbb{F} + \mathbb{G}\right)\overline{\mathbf{d}}^{n+1} + \mathbf{S}(\overline{\mathbf{d}}^{n+1}, \overline{\boldsymbol{\gamma}}_f^{n+1}) \\[2ex] = \rho_s\dfrac{1}{(\Delta t)^2}\mathbb{M}_3\overline{\mathbf{d}}^{II} + \dfrac{1}{\Delta t}\mathbb{F}\overline{\mathbf{d}}^I + \mathbf{p}^{endo}(t^{n+1}) - \mathbf{S}_0, \end{array}$$

$$(3.24)$$

by means of the Newton method. We notice that the vector $\overline{\boldsymbol{\gamma}}_f^{n+1}$ in Eq. (3.24) is already known, since it is given after the last step of Eq. (3.22) (i.e. for $m = N_{sub}$).

3.4.3 Partially Segregated Strategy $(\mathcal{I}_{SI}\mathcal{E}_{SI}\mathcal{A}_{SI})$–$(\mathcal{M}_I)$

By considering now the semi-implicit scheme (subscript SI) for the time discretization, the $(\mathcal{I}_{SI}\mathcal{E}_{SI}\mathcal{A}_{SI})$ problem reads:

$$(\mathcal{I}_{SI}\mathcal{E}_{SI}\mathcal{A}_{SI}): \begin{cases} \left(\dfrac{\vartheta_0^1}{\Delta t} + \mathbb{U}(\overline{\mathbf{v}}^*)\right)\overline{\mathbf{w}}^{n+\frac{m}{N_{sub}}} = \dfrac{1}{\tau}\overline{\mathbf{w}}^1 + \mathbf{Q}(\overline{\mathbf{v}}^*), \\[3mm] \left(\dfrac{\vartheta_0^1}{\Delta t}\mathbf{M} + \mathbf{K}(\overline{\mathbf{d}}^*) + \mathbb{I}_v^{ion}(\overline{\mathbf{v}}^*, \overline{\mathbf{w}}^*)\right)\overline{\mathbf{v}}^{n+\frac{m}{N_{sub}}} + \mathbb{I}_{\mathbf{w}}^{ion}(\overline{\mathbf{v}}^*, \overline{\mathbf{w}}^*)\overline{\mathbf{w}}^{n+\frac{m}{N_{sub}}} \\[3mm] \qquad\qquad = \dfrac{1}{\tau}\mathbf{M}\overline{\mathbf{v}}^1 + \widetilde{\mathbf{I}}^{ion}(\overline{\mathbf{v}}^*, \overline{\mathbf{w}}^*) + \mathbf{M}\mathbf{I}^{app}(t^{n+\frac{m}{N_{sub}}}), \\[3mm] \left(\dfrac{\vartheta_0^1}{\tau}\mathbf{M} + \varepsilon\mathbf{K}(\overline{\mathbf{w}}^*) + \mathbf{P}_{\gamma_f}(\overline{\mathbf{w}}^*, \overline{\boldsymbol{\gamma}}_f^*, \overline{\mathbf{d}}^*)\right)\overline{\boldsymbol{\gamma}}_f^{n+\frac{m}{N_{sub}}} = \dfrac{1}{\tau}\mathbf{M}\overline{\boldsymbol{\gamma}}_f^1 + \widetilde{\boldsymbol{\Phi}}(\overline{\mathbf{w}}^*, \overline{\boldsymbol{\gamma}}_f^*, \overline{\mathbf{d}}^*), \end{cases}$$
$$(3.25)$$

for $m = 1, \ldots, N_{sub}$. In this case, the block pattern of the matrix $\mathbb{A}_{EA}$, stemming from the linear system (3.25), is:

$$\mathbb{A}_{EA} = \begin{bmatrix} \boxed{\begin{matrix} A_{11} & 0 \\ A_{21} & A_{22} \end{matrix}} & \begin{matrix} 0 \\ 0 \end{matrix} \\ \begin{matrix} 0 & 0 \end{matrix} & \boxed{A_{33}} \end{bmatrix}.$$
$$(3.26)$$

As in the case of the $(\mathcal{I}_I\mathcal{E}_I\mathcal{A}_I)$ strategy, after solving Eq. (3.25) for N_{sub} steps, we solve the implicit mechanics problem $(\mathcal{M}_I)$ (3.24).

3.4.4 Fully Segregated Strategy $(\mathcal{I}_{SI})$–$(\mathcal{E}_{SI})$–$(\mathcal{A}_{SI})$–$(\mathcal{M}_I)$

Finally, we further segregate the $(\mathcal{I}_{SI}\mathcal{E}_{SI}\mathcal{A}_{SI})$ block: that is instead of solving schematic (3.25) in a single shot, we solve the three subproblems sequentially. In Fig. 3.4 we show a representation of the time advancement in this case. At each time t^n, the algorithm amounts to perform, for $m = 1, \ldots, N_{sub}$, the following steps, in order:

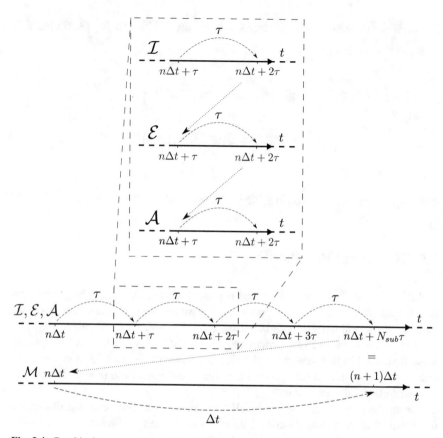

Fig. 3.4 Graphical representation of the time advancement for $(\mathcal{I}_{SI})$–$(\mathcal{E}_{SI})$–$(\mathcal{A}_{SI})$–$(\mathcal{M}_{I})$

1. find $\overline{\mathbf{w}}^{n+\frac{m}{N_{sub}}}$ by solving:

$$(\mathcal{I}_{SI}): \qquad \left(\frac{\vartheta_0^{\mathrm{I}}}{\Delta t} + \mathbb{U}(\overline{\mathbf{v}}^*)\right)\overline{\mathbf{w}}^{n+\frac{m}{N_{sub}}} = \frac{1}{\tau}\overline{\mathbf{w}}^{\mathrm{I}} + \mathbf{Q}(\overline{\mathbf{v}}^*); \qquad (3.27)$$

2. use $\overline{\mathbf{w}}^{n+\frac{m}{N_{sub}}}$, obtained with Eq. (3.27), to find $\overline{\mathbf{v}}^{n+\frac{m}{N_{sub}}}$ by solving:

$$\left(\frac{\vartheta_0^{\mathrm{I}}}{\Delta t}\mathbb{M} + \mathbb{K}(\overline{\mathbf{d}}^*) + \mathbb{I}_v^{ion}(\overline{\mathbf{v}}^*, \overline{\mathbf{w}}^{n+\frac{m}{N_{sub}}})\right)\overline{\mathbf{v}}^{n+\frac{m}{N_{sub}}}$$

$$(\mathcal{E}_{SI}):$$

$$= \frac{1}{\tau}\mathbb{M}\overline{\mathbf{v}}^{\mathrm{I}} + \widetilde{\mathbf{I}}^{ion}(\overline{\mathbf{v}}^*, \overline{\mathbf{w}}^{n+\frac{m}{N_{sub}}}) - \mathbb{I}_{\mathbf{w}}^{ion}(\overline{\mathbf{v}}^*, \overline{\mathbf{w}}^{n+\frac{m}{N_{sub}}})\overline{\mathbf{w}}^{n+\frac{m}{N_{sub}}} + \mathbb{M}\mathbf{I}^{app}(t^{n+\frac{m}{N_{sub}}}); \qquad (3.28)$$

3. use $\overline{\mathbf{w}}^{n+\frac{m}{N_{sub}}}$ and $\overline{\mathbf{v}}^{n+\frac{m}{N_{sub}}}$, obtained from Eq. (3.27) and Eq. (3.28), to find $\overline{\boldsymbol{\gamma}}_f^{n+\frac{m}{N_{sub}}}$ by solving:

$$(\mathcal{A}_{SI}): \quad \left(\frac{\vartheta_0^{\mathrm{I}}}{\tau} \mathbb{M} + \varepsilon \mathbb{K}(\overline{\mathbf{w}}^{n+\frac{m}{N_{sub}}}) + \mathbb{P}^{\gamma_f}(\overline{\mathbf{w}}^{n+\frac{m}{N_{sub}}}, \overline{\boldsymbol{\gamma}}_f^*, \overline{\mathbf{d}}^*) \right) \overline{\boldsymbol{\gamma}}_f^{n+\frac{m}{N_{sub}}}$$

$$= \frac{1}{\tau} \mathbb{M} \overline{\boldsymbol{\gamma}}_f^{\mathrm{I}} + \widetilde{\boldsymbol{\Phi}}(\overline{\mathbf{w}}^{n+\frac{m}{N_{sub}}}, \overline{\boldsymbol{\gamma}}_f^*, \overline{\mathbf{d}}^*).$$

$$(3.29)$$

After N_{sub} steps, we solve once again problem (3.24) and finally obtain $\overline{\mathbf{d}}^{n+1}$.

3.5 Numerical Results

In this section we first briefly describe the procedures used to obtain the geometries, the fiber and sheet fields, and the prestress, then we test the three segregated schemes on benchmark problems in both idealized and subject-specific LV geometries. The idealized mesh features 1827 vertices and 6500 tetrahedra, while the subject-specific mesh features 126,031 vertices and 637,379 tetrahedra. We use finite elements of order $r = 1$ and BDF of order $\sigma = 1$ (i.e. Backward Euler) and $\sigma = 2$ for the time discretization to ensure A-stability [63].

For all the simulations we use LifeV,[1] an open-source finite element library for the solution of problems described by PDEs in a High Performance Computing framework.

3.5.1 Preprocessing

Image segmentation locates regions of interest (ROI) in the form of a subset of pixels [35], and amounts to assign different flags to regions containing different types of tissues and/or fluids. In this work, to obtain the subject-specific mesh, we used a manual procedure exploiting the brightness of the pixels of a 3D MRI image;[2] see Fig. 3.5.

Fibers and sheets field distributions in the myocardium are not tipically extracted from MRIs, unless special procedures are applied [65]. For this reason, several

---

[1] https://cmcsforge.epfl.ch.

[2] The MRI images are provided by Prof. J. Schwitter (Chief physician at the Centre Hospitalier Universitaire Vaudois CHUV, Lausanne) and Dr. P. Masci (CHUV) in the framework of the collaboration CMCS@EPFL–CHUV.

Fig. 3.5 From left to right: the MRI from which the subject-specific geometry was segmented, the mesh, the fibers field, and the sheets field

mathematically rule-based definition of these fields have been used [33, 45, 50, 70], which attempt to construct their orientation. At the epicardium and at the endocardium, the fiber direction $\mathbf{f}_0$ is tangential to the boundary, while the sheet direction s_0 belongs to the plane identified by the normal and the ventricle centerline (i.e. the line passing through the center of the ventricle). In the most general case, angles α_{endo}, α_{epi}, β_{endo}, and β_{epi}, representing the inclination of the fibers and the sheets with respect to the base plane, are assigned. The direction of fibers and sheets inside the myocardium is determined by a transmurally linear mapping. Here we consider the rule-based algorithm proposed in [87] and further developed in [68]; we set $\alpha_{endo} = -60°$, $\alpha_{epi} = +60°$, $\beta_{endo} = \beta_{epi} = 0°$. In Fig. 3.5, we show the fields obtained by applying the algorithm to the subject-specific mesh.

Regarding the prestress, we solve problem (3.12) by means of a continuation method. More precisely, given the EDP value p_{EDP}^{endo} and an integer S representing the number of steps, we first define a pressure ramp increment in the form: $p_k = \frac{k}{S} p_{EDP}^{endo}$, $k = 1, \ldots, S$. For each $k = 1, \ldots, S$, we set $p = p_k$ in Eq. (3.4) and solve the nonlinear system (in $\mathbf{d}$) by using the Newton method to obtain an increasingly accurate approximation of $\mathbf{P}_{0,k}$, i.e. the prestress corresponding to the pressure p_k. We refer the reader to [28] for more details.

3.5.2 Benchmark Problem with Idealized Geometry

In order to assess the properties of the proposed segregated schemes and to evaluate their behavior for different timestep sizes, we set up and solve a benchmark problem by using the idealized geometry. The contraction is triggered by applying a current in three distinct points at the endocardium while keeping the pressure at the EDP value $p_{EDP}^{endo} = 10$ mmHg. We choose this setting so that the volume of the idealized LV halves during the simulation, from the initial value of approximately 136 mL to around 68 mL, thus attaining a deformation which is comparable with physiological data. We set $T = 0.1$ s and $\tau = 1, 2, 4, 8, 12, 16, 24, 32 \times 10^{-5}$ s, with $N_{sub} = 1, 2, 4, 8, 16$ for the segregated strategies (being $\Delta t = N_{sub}\tau$ the timestep size for the mechanics), while $\Delta t = \tau$ for the monolithic strategy. The absolute

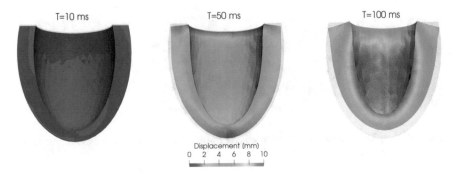

Fig. 3.6 Idealized LV and magnitude of the displacement field at different times, compared with the reference domain Ω_0, for the benchmark problem

tolerances for the Newton method and the GMRES solver are set to $\varepsilon_{tol}^N = 10^{-4}$ and $\varepsilon_{tol}^G = 10^{-8}$, respectively. For all the numerical simulations of this benchmark, 6 cpus are used.

In the following, we denote by $\widehat{v}_h^\tau$ and $\widehat{\mathbf{d}}_h^\tau$ the potential and the displacement solutions, respectively, obtained with $(\mathcal{I}_1\mathcal{E}_1\mathcal{A}_1\mathcal{M}_1)$ and timestep τ, while we set $\widehat{\tau} = 10^{-5}$ s (the smallest timestep size used). We use the solution generated by the monolithic approach $(\mathcal{I}_1\mathcal{E}_1\mathcal{A}_1\mathcal{M}_1)$ with $(\tau = \widehat{\tau})$ as a reference one (a manufactured "exact" solution). Indeed, since no exact solution for the electromechanics problem is available, the error analysis with respect to the timesteps is carried out by using a reference solution on the same mesh, effectively disregarding the error due to the space discretization. In the monolithic case, all the coupling conditions between the core models are enforced in the extradiagonal blocks of the monolithic system matrix. However, as we will show, this accuracy comes at the price of a large computational cost. In Fig. 3.6 we report the deformation and the displacement field of the ideal geometry obtained by solving the problem with $(\mathcal{I}_1\mathcal{E}_1\mathcal{A}_1\mathcal{M}_1)$ compared with the reference domain Ω_0.

We first numerically verify that in the $(\mathcal{I}_1\mathcal{E}_1\mathcal{A}_1\mathcal{M}_1)$ case the errors in $L^2(\Omega_0)$ and $L^\infty(\Omega_0)$ norms of the potential and of the displacement magnitude decay as τ and τ^2 when using BDF with $\sigma = 1$ and $\sigma = 2$, respectively. With this aim, we display in Fig. 3.7 the errors $||\widehat{v}_h^\tau - \widehat{v}_h^{\widehat{\tau}}||$ and $||\widehat{\mathbf{d}}_h^\tau - \widehat{\mathbf{d}}_h^{\widehat{\tau}}||$ against the timestep $\tau = \Delta t$. The converge rate is indeed coherent with the order σ of the BDF scheme under use.

Before analyzing the convergence rates against τ for the segregated schemes, we investigate the role of the splitting scheme on the mechanical feedback in the monodomain equation, which is realized by the dependence of the diffusion tensor on the deformation gradient $\mathbf{F}$. To this aim, we display in Fig. 3.8 the errors, at times $T = 24, 48, 96$ ms, made using $(\mathcal{I}_1\mathcal{E}_1\mathcal{A}_1)$–$(\mathcal{M}_1)$ with $\widehat{\tau}$ for different values of the parameter N_{sub}. That is, we use $\tau = \widehat{\tau}$ for the electrophysiology and the activation— the same timestep used to obtain the reference solution—while using $\Delta t = N_{sub}\tau$ for the mechanics. We first of all observe that, in all cases, the convergence rate is linear with respect to N_{sub} (equivalently, with respect to Δt); this behavior is expected since the Godunov splitting scheme used in the segregated algorithm is

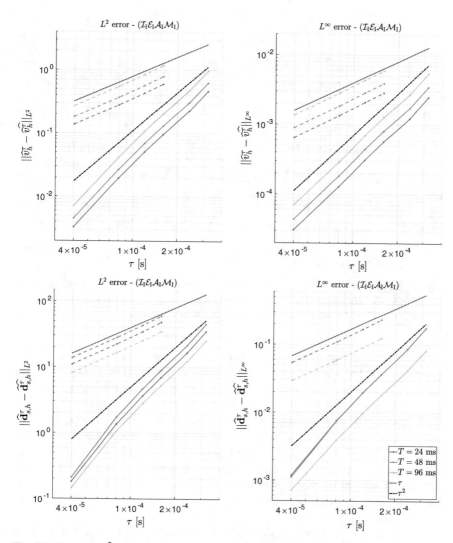

Fig. 3.7 Errors in $L^2(\Omega_0)$ (left) and $L^\infty(\Omega_0)$ (right) norms of the potentials $\widehat{v}_h^\tau$ (top) and of the displacements $\widehat{\mathbf{d}}_h^\tau$ (bottom) obtained by solving the problem with the monolithic scheme $(\mathcal{I}_I\mathcal{E}_I\mathcal{A}_I\mathcal{M}_I)$, at times $T = 24$ ms (blue), $T = 48$ ms (red), and $T = 96$ ms (yellow), in logarithmic scale against τ. The results for both BDF1 and BDF2 are reported

first order accurate. Furthermore, while the magnitude of the error of the potential $\widehat{v}_h^\tau$ is negligible when compared to the errors in Fig. 3.7, the same does not hold for the displacement $\widehat{\mathbf{d}}_h^\tau$. This is also expected since, in Fig. 3.8, the value of τ is fixed while Δt is not, nonetheless this clearly shows that, for transmembrane potential, the splitting error is several orders of magnitude smaller than the one introduced when using a larger τ.

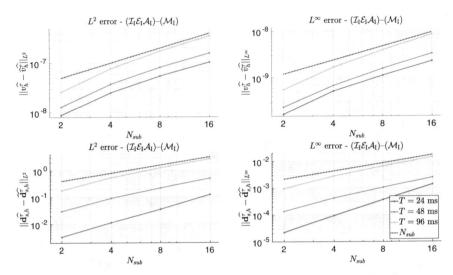

Fig. 3.8 Errors in $L^2(\Omega_0)$ (left) and $L^\infty(\Omega_0)$ (right) norms of the potentials $v_h^{\widehat{\tau}}$ (top) and of the displacements $\mathbf{d}_h^{\widehat{\tau}}$ (bottom) at time $T = 96$ ms obtained by solving the problem with the segregated scheme $(\mathcal{I}_I\mathcal{E}_I\mathcal{A}_I)$–$(\mathcal{M}_I)$ and timestep τ, in logarithmic scale against N_{sub}

We then analyze the errors introduced when using $(\mathcal{I}_I\mathcal{E}_I\mathcal{A}_I)$–$(\mathcal{M}_I)$, $(\mathcal{I}_{SI}\mathcal{E}_{SI}\mathcal{A}_{SI})$–$(\mathcal{M}_I)$, and $(\mathcal{I}_{SI})$–$(\mathcal{E}_{SI})$–$(\mathcal{A}_{SI})$–$(\mathcal{M}_I)$ for varying τ and N_{sub}. We do not report the ones for the case $N_{sub} = 1$ against τ (as previously done for the monolithic strategy) neither for $(\mathcal{I}_I\mathcal{E}_I\mathcal{A}_I)$–$(\mathcal{M}_I)$ nor for $(\mathcal{I}_{SI})$–$(\mathcal{E}_{SI})$–$(\mathcal{A}_{SI})$–$(\mathcal{M}_I)$ since no appreciable differences are visible with respect to $(\mathcal{I}_I\mathcal{E}_I\mathcal{A}_I\mathcal{M}_I)$ and $(\mathcal{I}_{SI}\mathcal{E}_{SI}\mathcal{A}_{SI})$–$(\mathcal{M}_I)$, respectively. In Fig. 3.9 the errors for the scheme $(\mathcal{I}_I\mathcal{E}_I\mathcal{A}_I)$–$(\mathcal{M}_I)$ at time $T = 96$ ms are reported for different choices of N_{sub}. As previously mentioned, only first order accuracy is granted by the considered splitting schemes, however the error on the potential v_h^τ converges even quadratically (it is superconvergent). On the other hand, the error on the displacement $\mathbf{d}_h^\tau$ converges linearly, unless the values $N_{sub} = 1, 2$ are employed.

By considering now the $(\mathcal{I}_{SI}\mathcal{E}_{SI}\mathcal{A}_{SI})$–$(\mathcal{M}_I)$ strategy, we report in Fig. 3.10 the errors made for the potential and the displacement at times $T = 24, 48, 96$ ms. We observe that the error significantly increases for $\tau > 8 \times 10^{-5}$ s, in particular for the potential v_h^τ. This is due to numerical instabilities occuring when using the semi-implicit case, which is "not guaranteed" to be stable for an arbitrary choice of τ. Nonetheless, these instabilities are "non-destructive" since the error on $\mathbf{d}_h^\tau$ is not significantly affected by them. However, in both cases, we observe again that the errors are superconvergent for $N_{sub} = 1$ and $\tau \le 8 \times 10^{-5}$ s as they decrease quadratically. We conclude our error analysis by reporting in Fig. 3.11 the errors against Δt at time $T = 96$ ms when using $(\mathcal{I}_{SI}\mathcal{E}_{SI}\mathcal{A}_{SI})$–$(\mathcal{M}_I)$, for different choices of N_{sub}. We conclude that, similarly to the $(\mathcal{I}_I\mathcal{E}_I\mathcal{A}_I)$–$(\mathcal{M}_I)$ case, the errors converge at least linearly, as expected by the splitting schemes employed.

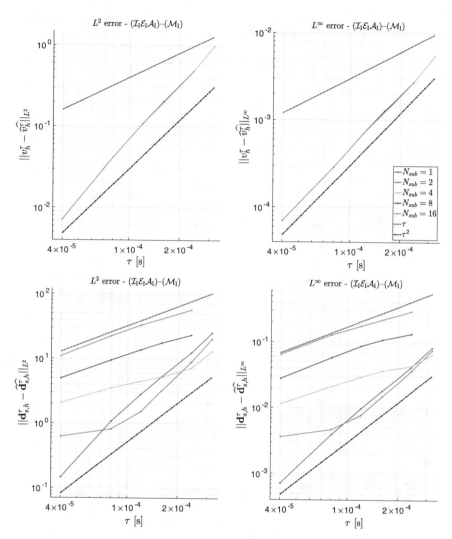

Fig. 3.9 Errors in $L^2(\Omega_0)$ (left) and $L^\infty(\Omega_0)$ (right) norms of the potentials v_h^τ (top) and of the displacements $\mathbf{d}_h^\tau$ (bottom) at time $T = 96\,\mathrm{ms}$ obtained by solving the problem with $(\mathcal{I}_I\mathcal{E}_I\mathcal{A}_I)$–$(\mathcal{M}_I)$ and $N_{sub} = 1, 2, 4, 8, 16$, in logarithmic scale against τ

We now investigate the efficiency of the schemes as a function of τ and N_{sub}. In Tables 3.1 and 3.2 we report the average number of Newton $(\overline{N}^N)$ and GMRES $(\overline{N}^N)$ iterations required for the solution of the monolithic problem with $(\mathcal{I}_I\mathcal{E}_I\mathcal{A}_I\mathcal{M}_I)$, and for the solution of the mechanics problem with the $(\mathcal{I}_I\mathcal{E}_I\mathcal{A}_I)$–$(\mathcal{M}_I)$, $(\mathcal{I}_{SI}\mathcal{E}_{SI}\mathcal{A}_{SI})$–$(\mathcal{M}_I)$, and $(\mathcal{I}_{SI})$–$(\mathcal{E}_{SI})$–$(\mathcal{A}_{SI})$–$(\mathcal{M}_I)$ strategies. As expected, both $\overline{N}^N$ and $\overline{N}^G$ increase significantly as τ gets larger and larger. Otherwise, the required wall time T^W dramatically drops with a speed-up of

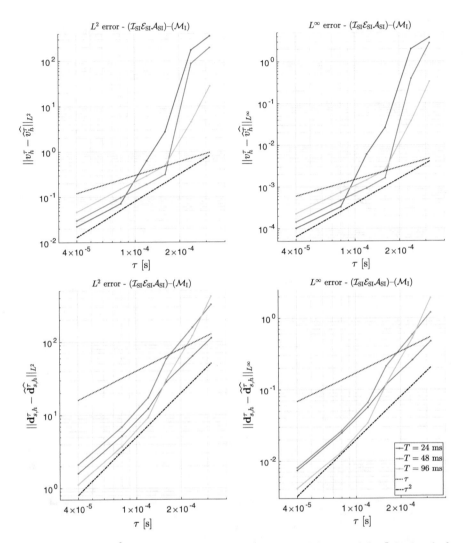

Fig. 3.10 Errors in $L^2(\Omega_0)$ (left) and $L^\infty(\Omega_0)$ (right) norms of the potentials v_h^τ (top) and of the displacements $\mathbf{d}_h^\tau$ (bottom) obtained by solving the problem with $(\mathcal{I}_{SI}\mathcal{E}_{SI}\mathcal{A}_{SI})$–$(\mathcal{M}_I)$, at times $T = 24$ ms (blue), $T = 48$ ms (red), and $T = 96$ ms (yellow), in logarithmic scale against τ, for $N_{sub} = 1$

almost 300 times when using the strategy $(\mathcal{I}_{SI})$–$(\mathcal{E}_{SI})$–$(\mathcal{A}_{SI})$–$(\mathcal{M}_I)$ and timestep $\tau = 32 \times 10^{-5}$, with respect to $(\mathcal{I}_I\mathcal{E}_I\mathcal{A}_I\mathcal{M}_I)$ with timestep $\tau = 1 \times 10^{-5}$.

We can now better compare the efficiency of the different strategies by displaying in Fig. 3.12 the errors of the displacement in $L^2(\Omega)$ norm against the total wall time T^W, and hence establish which strategy is the most efficient for a given tolerance on the error. The first clear conclusion that can be drawn is that, whatever the tolerance, it is more convenient to use the $(\mathcal{I}_I\mathcal{E}_I\mathcal{A}_I)$–$(\mathcal{M}_I)$ strategy with $N_{sub} = 1$ rather than

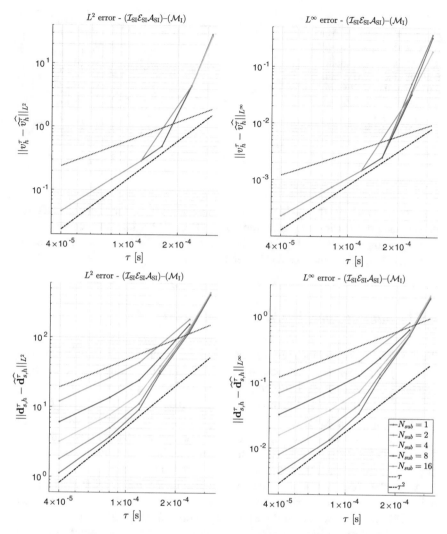

Fig. 3.11 Errors in $L^2(\Omega_0)$ (left) and $L^\infty(\Omega_0)$ (right) norms of the potentials v_h^τ (top) and of the displacements $\mathbf{d}_h^\tau$ (bottom) at time $T = 96$ ms obtained by solving the problem with $(\mathcal{I}_{SI}\mathcal{E}_{SI}\mathcal{A}_{SI})$–$(\mathcal{M}_I)$ and $N_{sub} = 1, 2, 4, 8, 16$, in logarithmic scale against τ

the monolithic $(\mathcal{I}_I\mathcal{E}_I\mathcal{A}_I\mathcal{M}_I)$ one; this is in agreement with our previous observations on the magnitude of the splitting error introduced by using the $(\mathcal{I}_I\mathcal{E}_I\mathcal{A}_I)$–$(\mathcal{M}_I)$ strategy. More in general, we observe that the proposed segregated strategies represent a better alternative with respect to the monolithic one if a larger error on the displacement is deemed to be acceptable. We conclude that the chosen strategy represents a trade-off between the efficiency of the simulation and the accuracy of the approximated solution. If the goal is that of reducing the computational

Table 3.1 The average number of Newton ($\overline{N}^N$) and GMRES ($\overline{N}^G$) iterations for the solution of the monolithic problem ($\mathcal{I}_1\mathcal{E}_1\mathcal{A}_1\mathcal{M}_1$), and the total wall time (T^W, in minutes) for the benchmark simulations, for each $\tau\ (=\Delta t)$ considered

$\tau\ (=\Delta t)$	$(\mathcal{I}_1\mathcal{E}_1\mathcal{A}_1\mathcal{M}_1)$		
	$\overline{N}^N$	$\overline{N}^G$	T^W
10^{-5} s	2.1	8.5	363'
2×10^{-5} s	2.4	8.1	188'
4×10^{-5} s	3.0	7.8	101'
8×10^{-5} s	3.5	7.7	54'
12×10^{-5} s	3.6	7.9	36'
16×10^{-5} s	4.0	7.6	28'
24×10^{-5} s	4.7	7.9	22'
32×10^{-5} s	5.2	9.2	18'

cost, the $(\mathcal{I}_{SI})$–$(\mathcal{E}_{SI})$–$(\mathcal{A}_{SI})$–$(\mathcal{M}_1)$ strategy should be the matter of choice, although its accuracy drops for larger timestep sizes. On the other hand, if accuracy is the driving factor, $(\mathcal{I}_1\mathcal{E}_1\mathcal{A}_1)$–$(\mathcal{M}_1)$ has to be preferred to $(\mathcal{I}_1\mathcal{E}_1\mathcal{A}_1\mathcal{M}_1)$ thus avoiding the extremely long wall times needed by the latter.

3.5.3 Subject-Specific LV: The Full Heartbeat

We use the subject-specific mesh of Fig. 3.5 for the simulation of a full heartbeat by modeling the pressure as detailed in Sect. 3.3.2.1 and by setting $T = 0.8$ s. Basis functions of degree $r = 1$ are employed, thus obtaining a system of size $M = 8 \times N_1^{dof} = 1{,}008{,}248$ in the monolithic case, together with BDF of order $\sigma = 2$. The timestep is set equal to $\tau = 5 \times 10^{-5}$ s while $\Delta t = N_{sub}\tau$ with $N_{sub} = 1, 5, 10$ for the segregated strategies. All the numerical simulations were carried out using Piz Daint, a Cray XC50/XC40 supercomputer installed at the Swiss National Supercomputing Center (CSCS),[3] and 72 cores were used for each simulation.[4]

As for the previous test, a current is applied at the endocardium at three distinct points for 3 ms in order to trigger the cardiac rythm. We show the results obtained in Fig. 3.13, where the transmembrane potential at times $T = 10, 20, 40$ ms is depicted together with the activation time (AT). The latter is defined, in each point, as the time at which the electric potential reaches a threshold value v^{thr} (we set in particular $v^{thr} = 10$ mV) [56, 81]. The activation time is in good agreement with experimental

---

[3]http://www.cscs.ch.

[4]Unfortunately the maximum wall time allowed on the Piz Daint supercomputer is set to 24 h, thus making it impossible to complete a simulation of a full heartbeat in all cases—most notably for the $(\mathcal{I}_1\mathcal{E}_1\mathcal{A}_1\mathcal{M}_1)$ strategy, which is the most computational demanding. We hence run two sets of simulations: in the first case, we set $T = 0.8$ s thus obtaining the pressure-volume loops of Fig. 3.16; in the second one, we set $T = 0.073$ s (the maximum time reachable in 24 h with the $(\mathcal{I}_1\mathcal{E}_1\mathcal{A}_1\mathcal{M}_1)$ strategy) thus obtaining the results of Fig. 3.16.

Table 3.2 The average number of Newton ($\overline{N}^N$) and GMRES ($\overline{N}^G$) iterations for the solution of the mechanics problem and the total wall time (T^W, in minutes) for the benchmark simulations, for each segregated strategy, τ (in 10^{-5} s), and N_{sub} considered

$(\mathcal{I}_I\mathcal{E}_I\mathcal{A}_I)-(\mathcal{M}_I)$

τ	\multicolumn{3}{c}{$N_{sub}=1$}			\multicolumn{3}{c}{$N_{sub}=2$}			\multicolumn{3}{c}{$N_{sub}=4$}			\multicolumn{3}{c}{$N_{sub}=8$}			\multicolumn{3}{c}{$N_{sub}=16$}		
	$\overline{N}^N$	$\overline{N}^G$	T^W	$\overline{N}^N$	$\overline{N}^G$	T^W	$\overline{N}^N$	$\overline{N}^G$	T^W	$\overline{N}^N$	$\overline{N}^G$	T^W	$\overline{N}^N$	$\overline{N}^G$	T^W
1	2.5	4.4	300'	2.7	4.7	221'	2.8	5.6	178'	3.3	5.4	156'	3.6	7.0	144'
2	2.7	4.7	144'	2.8	5.6	101'	3.3	5.4	81'	3.6	7.0	69'	4.1	10.3	63'
4	2.8	5.6	70'	3.3	5.4	50'	3.6	7.0	38'	4.1	10.2	32'	4.5	14.5	29'
8	3.3	5.4	38'	3.6	7.0	26'	4.1	10.3	20'	4.4	14.6	16'	5.2	18.7	15'
12	3.4	6.2	26'	3.8	8.8	18'	4.4	12.6	13'	4.8	17.2	11'	5.4	21.9	10'
16	3.6	7.0	20'	4.2	10.2	14'	4.4	14.6	10'	5.2	18.7	9'	5.5	24.7	8'
24	3.8	8.9	15'	4.4	12.6	10'	4.8	17.2	8'	5.4	21.9	6'	6.0	28.8	6'
32	4.2	10.3	12'	4.5	14.6	8'	5.3	18.7	6'	5.5	24.8	5'	6.0	33.2	5'

$(\mathcal{I}_{SI}\mathcal{E}_{SI}\mathcal{A}_{SI})-(\mathcal{M}_I)$

τ	\multicolumn{3}{c}{$N_{sub}=1$}			\multicolumn{3}{c}{$N_{sub}=2$}			\multicolumn{3}{c}{$N_{sub}=4$}			\multicolumn{3}{c}{$N_{sub}=8$}			\multicolumn{3}{c}{$N_{sub}=16$}		
	$\overline{N}^N$	$\overline{N}^G$	T^W	$\overline{N}^N$	$\overline{N}^G$	T^W	$\overline{N}^N$	$\overline{N}^G$	T^W	$\overline{N}^N$	$\overline{N}^G$	T^W	$\overline{N}^N$	$\overline{N}^G$	T^W
1	2.5	4.4	258'	2.7	4.7	179'	2.8	5.6	137'	3.3	5.4	115'	3.6	7.0	107'
2	2.7	4.7	123'	2.8	5.6	81'	3.3	5.4	59'	3.6	7.0	48'	4.1	10.2	42'
4	2.8	5.6	60'	3.3	5.4	39'	3.6	7.0	27'	4.2	10.2	21'	4.5	14.5	18'
8	3.3	5.4	32'	3.6	7.0	20'	4.1	10.2	14'	4.5	14.5	10'	5.2	18.6	8'
12	3.4	6.2	22'	3.9	8.8	13'	4.4	12.5	9'	4.8	17.1	7'	5.4	21.8	5'
16	3.5	7.0	16'	4.1	10.2	10'	4.4	14.5	7'	5.2	18.6	5'	5.5	24.6	4'
24	3.7	8.7	11'	4.3	12.3	7'	4.7	16.9	4'	5.3	21.5	3'	6.0	28.4	2'
32	3.8	9.9	8'	4.3	13.9	5'	4.8	18.5	3'	5.3	23.9	2'	10.6	23.9	2'

$(\mathcal{I}_{SI})-(\mathcal{E}_{SI})-(\mathcal{A}_{SI})-(\mathcal{M}_I)$

τ	\multicolumn{3}{c}{$N_{sub}=1$}			\multicolumn{3}{c}{$N_{sub}=2$}			\multicolumn{3}{c}{$N_{sub}=4$}			\multicolumn{3}{c}{$N_{sub}=8$}			\multicolumn{3}{c}{$N_{sub}=16$}		
	$\overline{N}^N$	$\overline{N}^G$	T^W	$\overline{N}^N$	$\overline{N}^G$	T^W	$\overline{N}^N$	$\overline{N}^G$	T^W	$\overline{N}^N$	$\overline{N}^G$	T^W	$\overline{N}^N$	$\overline{N}^G$	T^W
1	2.5	3.2	242'	2.7	3.6	159'	2.8	4.4	118'	3.3	4.4	97'	3.6	5.8	84'
2	2.7	3.6	114'	2.8	4.5	71'	3.3	4.5	50'	3.6	5.8	38'	4.1	8.5	32'
4	2.8	4.5	57'	3.3	4.5	36'	3.6	5.8	24'	4.2	8.5	18'	4.5	12.2	14'
8	3.3	4.5	31'	3.6	5.9	18'	4.2	8.6	12'	4.5	12.2	8'	5.2	15.9	7'
12	3.4	6.0	21'	3.8	8.0	13'	4.4	11.1	8'	4.8	15.2	6'	5.4	19.6	4'
16	3.5	7.0	16'	4.1	10.1	10'	4.4	14.3	6'	5.2	18.3	4'	5.5	24.3	3'
24	3.7	8.2	11'	4.3	11.6	6'	4.6	15.7	4'	5.4	20.2	3'	6.0	26.7	2'
32	3.8	8.5	8'	4.3	12.0	5'	4.8	15.9	3'	5.4	20.5	2'	9.8	28.7	1'

data obtained from healthy patients [10, 83], since the complete activation of the myocardium takes around 40 ms.

In Fig. 3.14 we show the displacement magnitude on the deformed myocardium Ω, compared with the reference geometry Ω_0, at the times $T = 100, 200, 300$ ms. A significant thickening of the myocardium walls takes place, which is in accordance with experimental observations [64]. The model, however, only produces a moderate

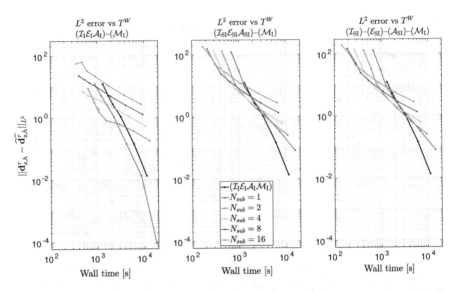

Fig. 3.12 Errors in $L^2(\Omega)$ norm of the displacement at time $T = 96$ ms against the total wall time T^W for the $(\mathcal{I}_1\mathcal{E}_1\mathcal{A}_1)$–$(\mathcal{M}_1)$, $(\mathcal{I}_{\rm SI}\mathcal{E}_{\rm SI}\mathcal{A}_{\rm SI})$–$(\mathcal{M}_1)$, and $(\mathcal{I}_{\rm SI})$–$(\mathcal{E}_{\rm SI})$–$(\mathcal{A}_{\rm SI})$–$(\mathcal{M}_1)$ strategies for $N_{sub} = 1, 2, 4, 8, 16$, compared to the error for the $(\mathcal{I}_1\mathcal{E}_1\mathcal{A}_1\mathcal{M}_1)$ strategy

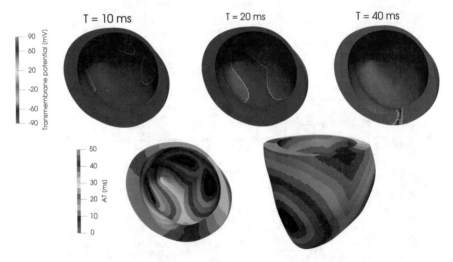

Fig. 3.13 Transmembrane potential at different times (top row) and activation time (bottom row) for the subject-specific simulation

rotation of the LV: in [58] the authors suggest that this behavior is related to the choice of the incompressibility constraint, the bulk modulus B magnitude, and to the boundary conditions. In order to better appreciate the behavior of the employed model, we also estimate the components of the stress tensor in the fibers and sheets direction $\sigma_{ff} = (\mathbf{P}\mathbf{f}_0)\mathbf{f}$ and $\sigma_{ss} = (\mathbf{P}\mathbf{s}_0)\mathbf{s}$. With this aim, we solve the following

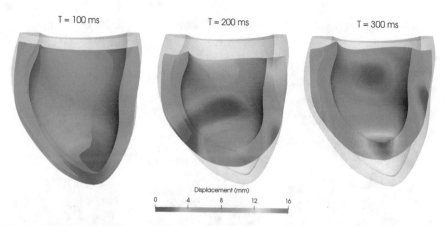

Fig. 3.14 Deformed subject-specific geometry and displacement field at different times, compared with the reference domain Ω_0, for the full heartbeat simulation

L^2-projection problem: find σ_{ff} such that

$$\int_{\Omega_0} \sigma_{ff} \psi_i = \int_{\Omega_0} (\mathbf{Pf}_0)\mathbf{f}\psi_i,$$

for $i, \ldots, N_1^{\text{dof}}$, and analogously for σ_{ss}. In Fig. 3.15 we show the two fields obtained at the same time instants considered in Fig. 3.14; we highlight that $T = 200$ ms corresponds approximately to the time at which the LV pressure attains its maximum (around 120 mmHg). The values assumed by σ_{ss} mostly fall in the physiological range [39, 76, 86, 88] and match the pressure value at the endocardium. Nonetheless the stress value peaks in the region close to the myocardium base; we believe that this is due to the thickness of the septum wall which, in this subject-specific geometry, was reconstructed as particularly thin. Regarding the stress σ_{ff}, the model reproduces much larger values with respect to those indicated in [6, 39, 88], thus overestimating them by almost an order of magnitude especially where the myocardium wall is (much) thinner. We remark, however, that the available medical data used in [39] to fit the strain energy function are obtained with in vitro loading tests, hence accounting only for the passive component of the stress.

Finally, we compare in Fig. 3.16 the pressure-volume (pV) loops obtained with the different numerical coupling strategies. A close up of the pV loops in the ejection phase is also reported to better assess the differences among them. We observe that, as in the benchmark test, the difference between the results obtained with $(\mathcal{I}_{\text{SI}}\mathcal{E}_{\text{SI}}\mathcal{A}_{\text{SI}})$–$(\mathcal{M}_{\text{I}})$ and those obtained with $(\mathcal{I}_{\text{SI}})$–$(\mathcal{E}_{\text{SI}})$–$(\mathcal{A}_{\text{SI}})$–$(\mathcal{M}_{\text{I}})$ is negligible. We conclude that the main deviation among the pV loops is caused by the choice of different timestep lengths $\Delta t = N_{sub}\tau$ for the mechanical model. Specifically, during the last part of the first isovolumic phase, the endocardial pressure increases very rapidly, while the change of the phase (as detailed in Sect. 3.2.4) takes place

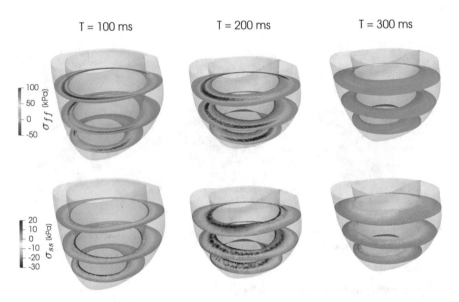

Fig. 3.15 Stress components σ_{ff} (top) and σ_{ss} (bottom) depicted on three slices of the deformed domain at three different times

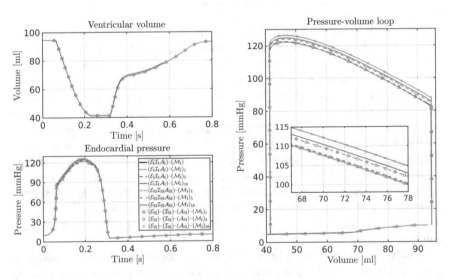

Fig. 3.16 LV internal volumes (top left) and endocardial pressures (bottom left) versus time, with pV loops (right) for the subject-specific simulations with all the strategies considered (the parameter N_{sub} used is indicated in legend with a subscript)

Table 3.3 The mechanics timestep Δt (in 10^{-5} s), the average number of Newton ($\overline{N}^N$) and GMRES ($\overline{N}^G$) iterations, and the total wall time (T^W, in minutes) for the simulation of the heartbeat with final time $T = 0.073$ s with the subject-specific mesh, using the four strategies considered and $N_{sub} = 1, 5, 10$

Strategy	N_{sub}	Δt	$\overline{N}^N$	$\overline{N}^G$	T^W
$(\mathcal{I}_I\mathcal{E}_I\mathcal{A}_I\mathcal{M}_I)$	–	5	3.3	18.7	1440'
$(\mathcal{I}_I\mathcal{E}_I\mathcal{A}_I)$–$(\mathcal{M}_I)$	1	10	4.4	40.8	723'
$(\mathcal{I}_I\mathcal{E}_I\mathcal{A}_I)$–$(\mathcal{M}_I)$	5	25	5.1	71.8	284'
$(\mathcal{I}_I\mathcal{E}_I\mathcal{A}_I)$–$(\mathcal{M}_I)$	10	50	5.8	93.6	259'
$(\mathcal{I}_{SI}\mathcal{E}_{SI}\mathcal{A}_{SI})$–$(\mathcal{M}_I)$	1	10	4.4	40.8	543'
$(\mathcal{I}_{SI}\mathcal{E}_{SI}\mathcal{A}_{SI})$–$(\mathcal{M}_I)$	5	25	5.0	72.1	136'
$(\mathcal{I}_{SI}\mathcal{E}_{SI}\mathcal{A}_{SI})$–$(\mathcal{M}_I)$	10	50	5.7	93.7	130'
$(\mathcal{I}_{SI})$–$(\mathcal{E}_{SI})$–$(\mathcal{A}_{SI})$–$(\mathcal{M}_I)$	1	10	4.4	37.7	582'
$(\mathcal{I}_{SI})$–$(\mathcal{E}_{SI})$–$(\mathcal{A}_{SI})$–$(\mathcal{M}_I)$	5	25	5.1	66.1	148'
$(\mathcal{I}_{SI})$–$(\mathcal{E}_{SI})$–$(\mathcal{A}_{SI})$–$(\mathcal{M}_I)$	10	50	5.7	86.4	93'

when the condition $p^{endo} \geq p^{ao}$ is satisfied. Hence, when using a large timestep Δt for the mechanical core model, the value of the pressure is higher when the ejection phase begins.

We conclude our analysis of the subject-specific simulations by reporting in Table 3.3 the values of $\overline{N}^N$, $\overline{N}^G$, and T^W, for the simulation of the heartbeat with final time set to $T = 0.073$ for all the strategies used. We observe that, even if in this case the number of Newton and GMRES iterations increases significantly with respect to the benchmark simulation case, the segregated schemes, and in particular the staggered schemes, allow to greatly reduce the computational costs for the subject-specific simulations too. Indeed, a speed-up of up to 16 times is obtained when using the $(\mathcal{I}_{SI})$–$(\mathcal{E}_{SI})$–$(\mathcal{A}_{SI})$–$(\mathcal{M}_I)$ strategy with $N_{sub} = 10$, compared to $(\mathcal{I}_I\mathcal{E}_I\mathcal{A}_I\mathcal{M}_I)$ used with the same timestep size τ.

3.6 Conclusions

We proposed several segregated strategies for the solution of the integrated electromechanics problem for the LV. We formulated the continuous model by coupling the monodomain equation, the ionic minimal model, the activation model for the fibers contraction, and the myocardial mechanics in the active strain framework. We approximated the mathematical model in space by means of the Finite Element method, and in time with both implicit and semi-implicit schemes based on BDF; then, we formulated segregated strategies for its solution, considering the more general case of staggered time discretizations arising from the choice of different timestep sizes for the electrophysiology and mechanical activation from one side, and for the mechanics from the other side.

The proposed segregated strategies were used with an idealized geometry for the simulation of a free contraction benchmark. The error on the results were evaluated against the solution obtained with $(\mathcal{I}_I\mathcal{E}_I\mathcal{A}_I\mathcal{M}_I)$ and $\tau = 10^{-5}$ s, here assumed to be almost "exact". We concluded from our error analysis that an approach based on segregating the mechanics from the rest of the problem allows to significantly reduce the computational cost of the simulation while providing accurate results: in the benchmark setting, it was possible to reduce the wall time for a 0.1 s long contraction approximately from 6 h with $(\mathcal{I}_I\mathcal{E}_I\mathcal{A}_I\mathcal{M}_I)$, to 2.5 h with $(\mathcal{I}_I\mathcal{E}_I\mathcal{A}_I)$–$(\mathcal{M}_I)$ and $N_{sub} = 16$. Using a semi-implicit time scheme for the electrophysiology and the activation allows to further cut the time required for the simulation, but at relatively large timesteps—for which the semi-implicit scheme is not stable—the accuracy significantly drops. Then, we showed that if also the electrophysiology and the activation are solved sequentially by segregating the ionic model, the monodomain equation and the mechanical activation, the computational cost is further reduced while the accuracy is preserved. With this approach, by using $(\mathcal{I}_{SI})$–$(\mathcal{E}_{SI})$–$(\mathcal{A}_{SI})$–$(\mathcal{M}_I)$ and $N_{sub} = 16$, we were able to solve the problem in less than 1.5 h.

Finally, we showed that the same integrated model can be used for large scale simulations with subject-specific geometries. We used the proposed strategies for the simulation of a full heartbeat and showed that physiological values for the pressure and the volume, are obtained. Segregated algorithms exhibit a significantly improved efficiency in this case too, when compared to the monolithic one. We conclude that segregated strategies are preferable if a relatively low temporal accuracy is acceptable, while the monolithic strategy should be preferred if the required accuracy is extremely high.

Acknowledgments This research was partially supported by the Swiss Platform for Advanced Scientific Computing (PASC, project "Integrative HPC Framework for Coupled Cardiac Simulations"). We also gratefully acknowledge the Swiss National Supercomputing Center (CSCS) for providing the CPU resources for the numerical simulations under project IDs s635/s796.

We acknowledge Prof. J. Schwitter and Dr. P. Masci (CHUV, Lausanne) for providing the MRI images used in this work and for the enlightening discussions. We also thank Prof. P. Tozzi (CHUV, Lausanne) for the invaluable insights in the functioning of the human heart.

The authors acknowledge the ERC Advanced Grant *iHEART*, "An Integrated Heart Model for the simulation of the cardiac function", 2017–2022, P.I. A. Quarteroni (ERC–2016–ADG, project ID: 740132).

References

1. R.R. Aliev, A.V. Panfilov, A simple two-variable model of cardiac excitation. Chaos, Solitons Fractals **7**(3), 293–301 (1996)
2. D. Ambrosi, S. Pezzuto, Active stress vs. active strain in mechanobiology: constitutive issues. J. Elast. **107**(2), 199–212 (2012)
3. D. Ambrosi, G. Arioli, F. Nobile, A. Quarteroni, Electromechanical coupling in cardiac dynamics: the active strain approach. SIAM J. Appl. Math. **71**(2), 605–621 (2011)
4. C.M. Augustin, A. Neic, M. Liebmann, A.J. Prassl, S.A. Niederer, G. Haase, G. Plank, Anatomically accurate high resolution modeling of human whole heart electromechanics: a strongly scalable algebraic multigrid solver method for nonlinear deformation. J. Comput. Phys. **305**, 622–646 (2016)

5. L. Barbarotta, A mathematical and numerical study of the left ventricular contraction based on the reconstruction of a patient specific geometry. Master thesis, Politecnico di Milano, Italy, 2014
6. L. Barbarotta, S. Rossi, L. Dedè, A. Quarteroni, A transmurally heterogeneous orthotropic activation model for ventricular contraction and its numerical validation. Int. J. Numer. Methods Biomed. Eng. **34**(12), e3137 (2018)
7. C. Brooks, H.H. Lu, *The Sinoatrial Pacemaker of the Heart* (Charles C. Thomas, Springfield Publisher, Springfield, 1972)
8. A. Bueno-Orovio, E.M. Cherry, F.H. Fenton, Minimal model for human ventricular action potentials in tissue. J. Theor. Biol. **253**(3), 544–560 (2008)
9. E. Burman, A. Ern, The discrete maximum principle for stabilized finite element methods, in *Numerical Mathematics and Advanced Applications* (Springer, Berlin, 2003), pp. 557–566
10. D.M. Cassidy, J.A. Vassallo, F.E. Marchlinski, A.E. Buxton, W.J. Untereker, M.E. Josephson, Endocardial mapping in humans in sinus rhythm with normal left ventricles: activation patterns and characteristics of electrograms. Circulation **70**(1), 37–42 (1984)
11. F.E. Cellier, E. Kofman, *Continuous System Simulation* (Springer, Berlin, 2006)
12. D. Chapelle, M. Fernández, J.F. Gerbeau, P. Moireau, J. Sainte-Marie, N. Zemzemi, Numerical simulation of the electromechanical activity of the heart, in *International Conference on Functional Imaging and Modeling of the Heart* (Springer, Berlin, 2009), pp. 357–365
13. P. Colli Franzone, L.F. Pavarino, G. Savaré, Computational electrocardiology: mathematical and numerical modeling, in *Complex Systems in Biomedicine* (Springer, Berlin, 2006), pp. 187–241
14. P. Colli Franzone, L.F. Pavarino, S. Scacchi, *Mathematical Cardiac Electrophysiology*, vol. 13 (Springer, Berlin, 2014)
15. P. Colli Franzone, L.F. Pavarino, S. Scacchi, Parallel multilevel solvers for the cardiac electro-mechanical coupling. Appl. Numer. Math. **95**, 140–153 (2015)
16. P. Colli Franzone, L.F. Pavarino, S. Scacchi, Joint influence of transmural heterogeneities and wall deformation on cardiac bioelectrical activity: a simulation study. Math. Biosci. **280**, 71–86 (2016)
17. P. Colli Franzone, L.F. Pavarino, S. Scacchi, Bioelectrical effects of mechanical feedbacks in a strongly coupled cardiac electro-mechanical model. Math. Models Methods Appl. Sci. **26**(1), 27–57 (2016)
18. P. Colli Franzone, L.F. Pavarino, S. Scacchi, Effects of mechanical feedback on the stability of cardiac scroll waves: a bidomain electro-mechanical simulation study. Chaos Interdisc. J. Nonlinear Sci. **27**(9), 093905 (2017)
19. F.S. Costabal, F.A Concha, D.E. Hurtado, E. Kuhl, The importance of mechano-electrical feedback and inertia in cardiac electromechanics. Comput. Methods Appl. Mech. Eng. **320**, 352–368 (2017)
20. H. Dal, S. Göktepe, M. Kaliske, E. Kuhl, A three-field, bi-domain based approach to the strongly coupled electromechanics of the heart. Proc. Appl. Math. Mech. **11**(1), 931–934 (2011)
21. H. Dal, S. Göktepe, M. Kaliske, E. Kuhl, A fully implicit finite element method for bidomain models of cardiac electromechanics. Comput. Methods Appl. Mech. Eng. **253**, 323–336 (2013)
22. M.J. Davis, J.A. Donovitz, J.D. Hood, Stretch-activated single-channel and whole cell currents in vascular smooth muscle cells. Am. J. Physiol. Cell Physiol. **262**(4), C1083–C1088 (1992)
23. S. Deparis, D. Forti, P. Gervasio, A. Quarteroni, INTERNODES: an accurate interpolation-based method for coupling the Galerkin solutions of PDEs on subdomains featuring non-conforming interfaces. Comput. Fluids **141**, 22–41 (2016)
24. S. Deparis, D. Forti, G. Grandperrin, A. Quarteroni, FaCSI: a block parallel preconditioner for fluid-structure interaction in hemodynamics. J. Comput. Phys. **327**, 700–718 (2016)
25. S. Doll, K. Schweizerhof, On the development of volumetric strain energy functions. J. Appl. Mech. **67**(1), 17–21 (2000)

26. T.S.E. Eriksson, A.J. Prassl, G. Plank, G.A. Holzapfel, Influence of myocardial fiber/sheet orientations on left ventricular mechanical contraction. Math. Mech. Solids **18**(6), 592–606 (2013)

27. D. Forti, M. Bukac, A. Quaini, S. Canic, S. Deparis, A monolithic approach to fluid–composite structure interaction. J. Sci. Comput. **72**(1), 396-421 (2017)

28. A. Gerbi, L. Dedè, A. Quarteroni, A monolithic algorithm for the simulation of cardiac electromechanics in the human left ventricle. Math. Eng. **1**(1), 1–37 (2018)

29. P. Gervasio, F. Saleri, A. Veneziani, Algebraic fractional–step schemes with spectral methods for the incompressible Navier–Stokes equations. J. Comput. Phys. **214**(1), 347–365 (2006)

30. G. Giantesio, A. Musesti, A continuum model of skeletal muscle tissue with loss of activation, in *Multiscale Models in Mechano and Tumor Biology*, ed. by A. Gerisch, R. Penta, J. Lang. Lecture Notes in Computational Science and Engineering, vol. 122 (Springer, Cham, 2017)

31. G. Giantesio, A. Musesti, On the modeling of internal parameters in hyperelastic biological materials (2017). arXiv preprint arXiv:1609.08651

32. S.K. Godunov, A difference method for numerical calculation of discontinuous solutions of the equations of hydrodynamics. Matematicheskii Sb. **89**(3), 271–306 (1959)

33. S. Göktepe, E. Kuhl, Electromechanics of the heart: a unified approach to the strongly coupled excitation–contraction problem. Comput. Mech. **45**(2–3), 227–243 (2010)

34. A.M. Gordon, A.F. Huxley, F.J. Julian, The variation in isometric tension with sarcomere length in vertebrate muscle fibres. J. Physiol. **184**(1), 170 (1966)

35. R.M. Haralick, L.G. Shapiro, Image segmentation techniques. Comput. Vis. Graphics Image Process. **29**(1), 100–132 (1985)

36. E.A. Heidenreich, J.M. Ferrero, M. Doblaré, J.F. Rodríguez, Adaptive macro finite elements for the numerical solution of monodomain equations in cardiac electrophysiology. Ann. Biomed. Eng. **38**(7), 2331–2345 (2010)

37. M. Hirschvogel, M. Bassilious, L. Jagschies, S.M. Wildhirt, M.W. Gee, A monolithic 3D–0D coupled closed-loop model of the heart and the vascular system: experiment-based parameter estimation for patient-specific cardiac mechanics. Int. J. Numer. Methods Biomed. Eng. **33**(8), e2842 (2017)

38. B.F. Hoffman, P.F. Cranefield, *Electrophysiology of the Heart* (McGraw-Hill, Blakiston Division, 1960)

39. G.A. Holzapfel, R.W. Ogden, Constitutive modelling of passive myocardium: a structurally based framework for material characterization. Philos. Trans. R. Soc. Lond. A Math. Phys. Eng. Sci. **367**(1902), 3445–3475 (2009)

40. M.C. Hsu, Y. Bazilevs, Blood vessel tissue prestress modeling for vascular fluid-structure interaction simulation. Finite Elem. Anal. Des. **47**(6), 593–599 (2011)

41. P.J. Hunter, M.P. Nash, G.B. Sands, Computational electromechanics of the heart. Comput. Biol. Heart **12**, 347–407 (1997)

42. A. Kamkin, I. Kiseleva, G. Isenberg, Stretch-activated currents in ventricular myocytes: amplitude and arrhythmogenic effects increase with hypertrophy. Cardiovasc. Res. **48**(3), 409–420 (2000)

43. A.M. Katz, *Physiology of the Heart* (Lippincott Williams & Wilkins, Philadelphia, 2010)

44. S. Land, S.A. Niederer, N.P. Smith, Efficient computational methods for strongly coupled cardiac electromechanics. IEEE Trans. Biomed. Eng. **59**(5), 1219–1228 (2012)

45. I. LeGrice, P. Hunter, A. Young, B. Smaill, The architecture of the heart: a data-based model. Philos. Trans. R. Soc. Lond. A Math. Phys. Eng. Sci. **359**(1783), 1217–1232 (2001)

46. J.R. Levick, *An Introduction to Cardiovascular Physiology* (Butterworth-Heinemann, Oxford, 2013)

47. C. Luo, Y. Rudy, A model of the ventricular cardiac action potential. Depolarization, repolarization, and their interaction. Circ. Res. **68**(6), 1501–1526 (1991)

48. C. Luo, Y. Rudy, A dynamic model of the cardiac ventricular action potential. I. Simulations of ionic currents and concentration changes. Circ. Res. **74**(6), 1071–1096 (1994)

49. M. Munteanu, L.F. Pavarino, Decoupled Schwarz algorithms for implicit discretizations of nonlinear monodomain and bidomain systems. Math. Models Methods Appl. Sci. **19**(7), 1065–1097 (2009)
50. D. Nickerson, N. Smith, P. Hunter, New developments in a strongly coupled cardiac electromechanical model. EP Europace **7**(2), 118–127 (2005)
51. F. Nobile, A. Quarteroni, R. Ruiz-Baier, An active strain electromechanical model for cardiac tissue. Int. J. Numer. Methods Biomed. Eng. **28**(1), 52–71 (2012).
52. D. Noble, A modification of the hodgkin–Huxley equations applicable to purkinje fibre action and pacemaker potentials. J. Physiol. **160**(2), 317–352 (1962)
53. D.A. Nordsletten, S.A. Niederer, M.P. Nash, P.J. Hunter, N.P. Smith, Coupling multi–physics models to cardiac mechanics. Prog. Biophys. Mol. Biol. **104**(1), 77–88 (2011)
54. R.W. Ogden, *Non-Linear Elastic Deformations* (Courier Corporation, North Chelmsford, 1997)
55. L.H. Opie, *Heart Physiology: From Cell to Circulation* (Lippincott Williams & Wilkins, Philadelphia, 2004)
56. S. Pagani, Reduced–order models for inverse problems and uncertainty quantification in cardiac electrophysiology. Ph.D. thesis, Politecnico di Milano, Italy, 2017
57. S. Pezzuto, Mechanics of the heart: constitutive issues and numerical experiments. Ph.D. thesis, Politecnico di Milano, Italy, 2013
58. S. Pezzuto, D. Ambrosi, Active contraction of the cardiac ventricle and distortion of the microstructural architecture. Int. J. Numer. Methods Biomed. Eng. **30**(12), 1578–1596 (2014)
59. M. Potse, B. Dubé, J. Richer, A. Vinet, R.M. Gulrajani, A comparison of monodomain and bidomain reaction-diffusion models for action potential propagation in the human heart. IEEE Trans. Biomed. Eng. **53**(12), 2425–2435 (2006)
60. A. Quarteroni, *Numerical Models for Differential Problems*. MS&A, vol. 16 (Springer, Berlin, 2017)
61. A. Quarteroni, L. Dede', A. Manzoni, C. Vergara, *Mathematical Modelling of the Human Cardiovascular System. Data, Numerical Approximation, Clinical Applications* (Cambridge University Press, Cambridge, 2019)
62. A. Quarteroni, T. Lassila, S. Rossi, R. Ruiz-Baier, Integrated heart—coupling multiscale and multiphysics models for the simulation of the cardiac function. Comput. Methods Appl. Mech. Eng. **314**, 345–407 (2017)
63. A. Quarteroni, R. Sacco, F. Saleri, *Numerical Mathematics*, vol. 37 (Springer, Berlin, 2010)
64. T.A. Quinn, P. Kohl, Combining wet and dry research: experience with model development for cardiac mechano-electric structure-function studies. Cardiovasc. Res. **97**(4), 601–611 (2013)
65. T.G. Reese, R.M. Weisskoff, R.N. Smith, B.R. Rosen, R.E. Dinsmore, V.J. Wedeen, Imaging myocardial fiber architecture in vivo with magnetic resonance. Magn. Reson. Med. **34**(6), 786–791 (1995)
66. F. Regazzoni, L. Dedè, A. Quarteroni, Active contraction of cardiac cells: a reduced model for sarcomere dynamics with cooperative interactions. Biomech. Model. Mechanobiol. **17**(6), 1663–1686 (2017)
67. B.M. Rocha, B. Lino, R.W. dos Santos, E.M. Toledo, L.P.S. Barra, J. Sundnes, A two dimensional model of coupled electromechanics in cardiac tissue, in *World Congress on Medical Physics and Biomedical Engineering*, Munich (Springer, Berlin, 2009), pp. 2081–2084
68. S. Rossi, Anisotropic modeling of cardiac mechanical activation. Ph.D. thesis, EPFL, Switzerland, 2014
69. S. Rossi, T. Lassila, R. Ruiz-Baier, A. Sequeira, A. Quarteroni, Thermodynamically consistent orthotropic activation model capturing ventricular systolic wall thickening in cardiac electromechanics. Eur. J. Mech. A Solids **48**, 129–142 (2014)
70. S. Rossi, R. Ruiz-Baier, L.F. Pavarino, A. Quarteroni, Orthotropic active strain models for the numerical simulation of cardiac biomechanics. Int. J. Numer. Methods Biomed. Eng. **28**(6–7), 761–788 (2012)

71. R. Ruiz-Baier, A. Gizzi, S. Rossi, C. Cherubini, A. Laadhari, S. Filippi, A. Quarteroni, Mathematical modelling of active contraction in isolated cardiomyocytes. Math. Med. Biol. **31**(3), 259–283 (2014)

72. Y. Saad, *Iterative Methods for Sparse Linear Systems* (SIAM, 2003)

73. P.P. Sengupta, J. Korinek, M. Belohlavek, J. Narula, M.A. Vannan, A. Jahangir, B.K. Khandheria, Left ventricular structure and function. J. Am. Coll. Cardiol. **48**(10) 1988–2001 (2006)

74. J.C. Simo, R.L. Taylor, Quasi-incompressible finite elasticity in principal stretches. Continuum basis and numerical algorithms. Comput. Methods Appl. Mech. Eng.**85**(3), 273–310 (1991)

75. R.J. Spiteri, R.C. Dean, On the performance of an Implicit–Explicit Runge–Kutta method in models of cardiac electrical activity. IEEE Trans. Biomed. Eng. **55**(5), 1488–1495 (2008)

76. D.D. Streeter, R.N. Vaishnav, D.J. Patel, H.M. Spotnitz, J. Ross, E.H. Sonnenblick, Stress distribution in the canine left ventricle during diastole and systole. Biophys. J. **10**(4), 345–363 (1970)

77. S. Sugiura, T. Washio, A. Hatano, J. Okada, H. Watanabe, T. Hisada, Multi-scale simulations of cardiac electrophysiology and mechanics using the university of Tokyo heart simulator. Prog. Biophys. Mol. Biol. **110**(2), 380–389 (2012)

78. K. Takizawa, Y. Bazilevs, T.E. Tezduyar, Space-time and ALE-VMS techniques for patient-specific cardiovascular fluid-structure interaction modeling. Arch. Comput. Methods Eng. **19**(2), 171–225 (2012)

79. K.H.W.J. Ten Tusscher, D. Noble, P.J. Noble, A.V. Panfilov, A model for human ventricular tissue. Am. J. Physiol. Heart Circ. Physiol. **286**(4), H1573–H1589 (2004)

80. N.A. Trayanova, Whole-heart modeling applications to cardiac electrophysiology and electromechanics. Circ. Res. **108**(1), 113–128 (2011)

81. T.P. Usyk, A.D. McCulloch, Electromechanical model of cardiac resynchronization in the dilated failing heart with left bundle branch block. J. Electrocardiol. **36**, 57–61 (2003)

82. T.P. Usyk, I.J. LeGrice, A.D. McCulloch, Computational model of three-dimensional cardiac electromechanics. Comput. Vis. Sci. **4**(4), 249–257 (2002)

83. J.A. Vassallo, D.M. Cassidy, J.M. Miller, A.E. Buxton, F.E. Marchlinski, M.E. Josephson, Left ventricular endocardial activation during right ventricular pacing: effect of underlying heart disease. J. Am. Coll. Cardiol. **7**(6), 1228–1233 (1986)

84. N. Westerhof, J.W. Lankhaar, B.E. Westerhof, The arterial Windkessel. Med. Biol. Eng. Comput. **47**(2), 131–141 (2009)

85. J.P. Whiteley, An efficient numerical technique for the solution of the monodomain and bidomain equations. IEEE Trans. Biomed. Eng. **53**(11), 2139–2147 (2006)

86. A.Y.K. Wong, P.M. Rautaharju, Stress distribution within the left ventricular wall approximated as a thick ellipsoidal shell. Am. J. **75**(5), 649–662 (1968)

87. J. Wong, E. Kuhl, Generating fibre orientation maps in human heart models using poisson interpolation. Comput. Methods Biomech. Biomed. Eng. **17**(11), 1217–1226 (2014)

88. F.C.P. Yin, R.K. Strumpf, P.H. Chew, S.L. Zeger Quantification of the mechanical properties of noncontracting canine myocardium under simultaneous biaxial loading. J. Biomech. **20**(6), 577–589 (1987)

Chapter 4
Power-Stroke-Driven Muscle Contraction

Raman Sheshka and Lev Truskinovsky

Abstract To show that acto-myosin contraction can be propelled directly through a conformational change, we present in these lecture notes a review of a recently developed approach to muscle contraction where myosin power-stroke is interpreted as the main active mechanism. By emphasizing the active role of power stroke, the proposed model contributes to building a conceptual bridge between processive and nonprocessive motors.

4.1 Introduction

Broadly accepted chemo-mechanical models of muscle contraction operate with kinetic constants depending on a continuous variable, the elongation of an effective spring. This leads to an appearance of phenomenological functions, which brings into the theory an infinite number of parameters. In an attempt to avoid such 'freedom' we make a tacit assumption that the structural elements of force producing machinery are purely mechanical and can be in principle built in a lab.

To model the phenomenon of muscle contraction at a purely mechanical level we follow the approach developed in the theory of Brownian ratchets which replaces the conventional chemistry-based interpretation of active force generation by the study of Langevin dynamics of mechanical systems. In ratchet-based models describing acto-myosin contraction, the ATP activity is usually associated with actin binding potential while the power-stroke mechanism, residing inside myosin heads, is viewed as passive. Instead, in view of the reasons discussed below, we assume that power-stroke is the main active mechanism. Implicitly, our basic assumption is that

R. Sheshka
NeoXam, Paris, France

L. Truskinovsky (✉)
ESPCI, PMMH, CNRS – UMR 7636 PSL-ESPCI, Paris, France
e-mail: lev.truskinovsky@espci.fr

© The Editor(s) (if applicable) and The Author(s), under exclusive licence
to Springer Nature Switzerland AG 2020
A. DeSimone et al., *The Mathematics of Mechanobiology*,
Lecture Notes in Mathematics 2260, https://doi.org/10.1007/978-3-030-45197-4_4

Fig. 4.1 Schematic
illustration of the four-step
Lymn–Taylor cycle showing
the power-stroke $A \rightarrow B$, the
detachment $B \rightarrow C$, the
re-cocking of the
power-stroke $C \rightarrow D$ and the
final re-attachment $D \rightarrow A$
bringing the system back into
the original state

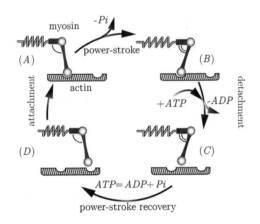

the fundamental mechanism behind active force generation is the same in the linear
molecular motors (myosins, kynesins, dyneines).

Muscle contraction takes the form of a relative motion of thick (myosin) and
thin (actin) filaments [3]. Active force generation results from stochastic interaction
between individual myosin heads (cross-bridges) and the adjacent actin binding
sites. It includes cyclic attachment of myosin cross-bridges to actin filaments
together with a concurrent conformational change in the core of the myosin
catalytic domain (of folding-unfolding type). A lever arm amplifies this structural
transformation producing the power-stroke, which allows the attached cross bridges
to generate macroscopic forces [53].

Myosin motors convert chemical energy into mechanical work by catalyzing the
hydrolysis of adenosin triphosphate (ATP) to the adenosine di-phosphate (ADP),
freeing phosphate (Pi) and using the released energy for generating motion. A
prototypical biochemical model of the myosin ATPase reaction in solution, linking
together the attachment-detachment, the power-stroke and the hydrolysis of ATP, is
known as the Lymn–Taylor cycle [72], see Fig. 4.1. While this minimal description
of enzyme kinetics is common for most myosins motors [113], its association
with microscopic structural details and its relation to micro-mechanical interactions
remains a subject of debate [22, 111, 112].

In physiological literature it is usually implied that muscle contraction is to
a large degree driven by the power-stroke which is then perceived as an active
mechanism [42]. This opinion is supported by observations that both the power-
stroke and the reverse-power-stroke can be induced by ATP even in the absence of
actin filaments [112], that contractions can be significantly inhibited by antibodies
which impair lever arm activity [84], that sliding velocity in mutational myosin
forms depends on the lever arm length [117] and that the directionality can be
reversed as a result of modifications in the lever arm domain [15, 116].

A perspective that the power-stroke is the driving force of active contraction
was challenged by the suggestion that myosin catalytic domain could operate as
a Brownian ratchet, which means that it can move and produce contraction without

assistance from the power-stroke mechanism [49, 95, 130]. In this interpretation the contraction is driven directly by the attachment-detachment machinery which rectifies correlated noise and selects directionality following, for instance, the polarity of actin filaments [57, 58].

Although the simplest models of Brownian ratchets neglect the conformational change, some phases of the attachment-detachment cycle can be interpreted as the power-stroke when the actin potential is assumed to undergoe additional externally driven horizontal shifts [122, 126]. Ratchet models were also proposed where the periodic spatial landscape is supplemented by a reaction coordinate, representing the conformational change [93, 127]. In all these models, however, the role of the power stroke was viewed as secondary and the contraction could be generated even if the power stroke mechanism was disabled. The main functionality of the power-stroke mechanism would be then attributed to passive fast force recovery [2, 19, 54].

An alternative viewpoint that the power-stroke mechanism consumes chemical energy and can be viewed as active, is the underpinning of the broadly accepted phenomenological chemo-mechanical models [52, 88]. These models pay great attention to structural details and in their most comprehensive versions faithfully reproduce the main experimental observations [83, 109]. The chemo-mechanical models, however, are not transparent mechanistically because they deal with elastic interactions implicitly. In these models chemical states are interpreted as continuous manifolds (parameterized by the strain) and to characterize jump processes between the points on these manifolds the authors choose the transition rate functions phenomenologically. While this functional freedom compensates the lack of knowledge of the underlying multidimensional energy landscape, the inherent arbitrariness of some of these choices limits the ultimate predictive power of this approach.

In an attempt to reach a synthetic description, several hybrid models, allowing chemical states to coexist with springs and forces, have been also proposed [26, 46, 65]. The phenomenological side of these models is minimal, however, they still combine continuous dynamics with jump transitions which makes the precise identification of structural prototypes and the underlying micro-mechanical interactions challenging. In this class of models the power-stroke in an individual cross-bridge was reproduced most faithfully by Geislinger and Kawai who introduced a 2D energy landscape by coupling a bi-stable potential with a symmetric periodic potential [43]. In this model both the attachment-detachment mechanism and the power stroke mechanism were effectively endowed with activity, however, the ATP hydrolysis was still represented by a flashing actin potential.

In these lecture notes we review a set of models which suggest that the power stroke can be, at least in principle, the main driving force behind muscle contraction. The discussed models are fully mechanistic in the sense that all ambiguous jump processes are replaced by a fully mechanistic Langevin dynamics. To focus exclusively on the idea of an active power stroke, driven directly by the ATP hydrolysis, we intentionally simplify the real picture and at some point even model actin filaments as passive non-polar tracks. The power-stroke mechanism is represented by a symmetric bi-stable potential associated with an internal degree of freedom and the ATP activity is modeled by a correlated component of the external noise. We show that even in this simple setting one can obtain a fully mechanical interpretation of all four chemical states in the minimal Lymn–Taylor cycle which opens a possibility of building artificial engineering devices mimicking enzymatic activity.

We start with a discussion of the three main force generation mechanisms: motor driven contraction, power stroke driven contraction and the model where power stroke not only drives the process but also controls the attachment/detachment mechanism. The idea that the symmetry breaking mechanism rests in internal conformational transition [23, 75] is borrowed from the theory of processive motors [24, 82, 104, 124]. Thus, in the description of dimeric motors it is usually assumed that ATP hydrolysis induces a conformational transformation which then changes the relative position of the motor legs ensuring motility [119]. Here we use the same idea to describe a non-processive motor with a single leg that remains on track due to the presence of a thick filament.

To be realistic, a model of the power stroke driven contraction must contain an assumption that the strength of the attachment depends on the state of the power stroke element. To justify the implied coupling, we argue that the conformational state of the power-stroke element provides steric regulation of the distance between the myosin head and the actin filament. More specifically, we assume that when the lever arm swings, the interaction of the head with the binding site weakens. This and other aspects of steric rotation-translation coupling in ratchet models are discussed in [43, 68, 90].

The proposed framework allows for three different modes of power stroke driven contractility which may operate simultaneously.

The first mode is activated only if correlations are present in the additive noise as in the conventional rocking ratchets [74]. The peculiarity of our rocking ratchet is that the periodic potential is symmetric and time independent. The correlated component of the noise affects the bi-stable potential and, since it is also symmetric, the directional motion is due exclusively to an asymmetry induced by the coupling between the internal degrees of freedom and the center of mass of the motor.

The second mode does not necessitate correlations in the noise but instead requires that the coupling between the power-stroke element and the actin filament is hysteretic. The motor can then extract energy directly from the delay mechanism which represents a non-equilibrium reservoir. We show that the two active mechanisms, correlations-induced and hysteresis-induced, can favor motions in different directions and can play complimentary roles.

Finally, the third mode functions if the internal forces acting between the myosin head and the actin filament are non-potential even without being hysteretic [73, 93, 130]. This assumption introduces another active mechanism which can drive the motor even if the environment is in equilibrium. The correlations-induced and non-potentiality-induced mechanisms can impose opposite directionality, in particular, they can operate in combination to slow down and even to stop the motor.

The variety of the available regimes is particularly rich when the forces are non-potential and the coupling between the power-stroke and the actin filament is hysteretic. The resulting ratchet shows complex reversals of current depending on the amplitude of the external driving and temperature. The importance of the hysteretic coupling is revealed by the observation that only in this case the model can reproduce all four steps of the Lymn–Taylor cycle.

In the last section of these lecture notes we show that the power-stroke-driven active rigidity can emerge as a result of resonant nonthermal excitation of the power stroke machinery. Building upon the idea of active drift [96] we consider a family of stall states in a power stroke driven system parameterized by a meso-scopic measure of total deformation. We compute the time and ensemble averaged potential at the fixed value of the deformation parameter and interpret the deformation derivative of this potential as an effective stiffness.

Of particular interest is the effect on active rigidity of the stochastic nature of the nonequilibrium reservoir. We show that while in periodic or dichotomous environments, the noise induced pseudo well exists, active stabilization disappears if the noise is of Ornshtein–Uhlenbeck type. The sensitive dependence of the mechanical performance of the molecular scale devices on the shape of the power spectrum of the noise, has been previously observed in the studies of active drift [25, 81] and here we broaden the picture by covering molecular machines generating active rigidity. Various features captured by our minimal model of active rigidity are in common not only with inertial stabilization [17], but also with the performance of the Ising model in periodic magnetic field [40], the folding/unfolding of proteins subjected to periodic forces [38] and the parametric behavior of more complex actively driven systems [11, 20, 85].

While some of the material collected in these lecture notes is new, the main ideas are based on the published papers [107, 108].

4.2 General Ratchet Model

Even the simplest mechanical model of a cross-bridge capable of both, the attachment detachment and the power stroke, must involve at least two continuous variables [52, 73, 74].

Suppose that the position of the motor head is modeled by the variable $x(t)$, while the variable $y(t)$ describes the internal degree of freedom representing the configuration of the lever-arm, see Fig 4.2a. Suppose also that the interaction between the myosin head and the actin filament is modeled by a space periodic potential. Finally, assume that the molecular link between the motor head and the motor tail can be described by a bi-stable spring, see Fig 4.2b.

Using these assumptions we can now formulate the general stochastic dynamics of a cross-bride in terms of the system of Langevin equations:

$$\eta d\mathbf{X}/dt = -\nabla G(\mathbf{X}, t) + \sigma \boldsymbol{\xi}(t). \tag{4.1}$$

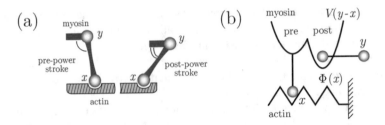

Fig. 4.2 (**a**) Sketch of one-legged molecular motor in pre-power stroke and post-power stroke configurations. (**b**) The mechanical model

Here η is a diagonal matrix defining drift coefficients, σ is another diagonal matrix with components $\sqrt{2\eta_i k_B \Theta}$, and $\xi(t)$ is the Gaussian random vector with zero mean $\langle \xi_i(t) \rangle = 0$, and with correlations $\langle \xi_i(t) \xi_j(s) \rangle = \delta_{ij} \delta(t - s)$, $i, j = x, y$. The terms $\eta_i dx_i / dt$ describe frictional forces and the corresponding drag coefficients are assumed to be constant. The function $G(\mathbf{X}, t)$ introduces the energy landscape of the motor device.

The mechanical action of the ATP hydrolysis will be represented by a correlated component of the external noise $\mathbf{f}$. Suppose that such a noise can affect both, the actin/myosin bound states and the conformational state of the lever-arm. This means that the corresponding force can act on coordinates x and y. Writing a generic expression for a tilted energy landscape $G(\mathbf{X}, t)$ we obtain

$$G(\mathbf{X}, t) = G_0(x, y) - x f_x(t) - y f_y(t). \tag{4.2}$$

Here $G_0(x, y) = \Phi(x) + V(y - x)$ is the intrinsic energy landscape, see Fig. 4.4a, where the bi-stable potential $V(y - x)$ describes two conformational states of the power stroke mechanism. We identify one energy well with the pre-power-stroke state and another energy well with the post-power-stroke state. We also assume that the potential $\Phi(x)$ has period L so what $\Phi(x + L) = \Phi(x)$. The simplest representation of the correlated noise $f_i(t)$ imitating the ATP activity is through periodic functions with zero average (over the corresponding periods τ_i).

Since our goal is to develop the model with active power stroke we will not consider the most general case, when the components $f_y(t)$ and $f_x(t)$ are independent. Instead, we focus on three specific models. In 'X model' the rocking/tilting force acts on the variable x,

$$f_x(t) = f(t), \quad f_y(t) = 0; \tag{4.3a}$$

this model was already introduced and studied in [76]. In 'Y model' the rocking force will be acting on the variable y

$$f_x(t) = 0, \quad f_y(t) = f(t). \tag{4.3b}$$

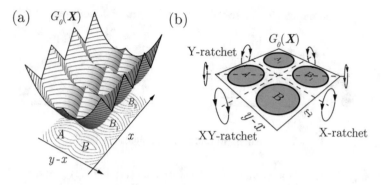

Fig. 4.3 (a) The surface and contour plot of the function $G_0(x, y)$. (b) The scheme illustrating three different rocking mechanism: X-ratchet (4.3a), Y-ratchet (4.3b), XY-ratchet (4.3c)

Finally, in 'XY model' the rocking couple will act on the difference $y - x$,

$$f_x(t) = -f(t), \quad f_y(t) = f(t). \tag{4.3c}$$

In Fig. 4.3a we illustrate our two-dimensional energy landscape identifying four different mechanical configurations A, B, A_1, B_1 which represent different local minima of the energy. The schematic illustration of the three different rocking mechanisms is shown in Fig. 4.3b.

In the X-tilted ratchet model (4.3a) the correlated force is applied to the x variable while the power-stroke mechanism remains passive. Applying the correlated force to the y variable in the Y-tilted ratchet model (4.3b) we make the first step in the direction of bringing activity to the power stroke mechanism. Finally, in the XY-tilted ratchet model (4.3c), the active force is applied directly to the $y - x$ variable. The non-equilibrium noise in this model is then acting directly on the internal degree of freedom characterizing the motor mechanism.

A model closely related to XY-tilted ratchet and coupling translational and rotational degrees of freedom was considered in [43]. The XY-tilted ratchet also resembles some models of Kinesin where ATP acts on the internal bi-stable device forcing two legs to move along the actin filament. In this sense the XY-tilted cross-bridge can be viewed as a "single-leg" analog of a Kinesin motor [78] with both mechanisms driven through the power stroke.

In these lecture notes we consider two loading devices–hard and soft, see Fig. 4.4. In the case of hard (or, rather, mixed) device the total energy is

$$G_0^h(x, y) = \Phi(x) + V(y - x) + \frac{1}{2}k_m(y - z)^2, \tag{4.4}$$

where $V_m = \frac{1}{2}k_m(y - z)^2$ is the potential of a linear spring and the variable z is treated as a fixed external parameter, see Fig. 4.4a. In this case, it is of interest to compute the average tension generated by such system $T = k_m(\langle y \rangle - z)$, where the

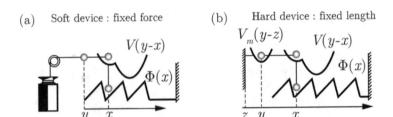

Fig. 4.4 (**a**) Soft device configuration; (**b**) hard device configuration

notation $\overline{\langle y \rangle}$ indicates the averaging of the stochastic variable $y(t)$ over the ensemble and over the time. In the case of soft device the total energy is

$$G_0^s(x, y) = \Phi(x) + V(y - x) - y f_{ext}, \tag{4.5}$$

where f_{ext} is an external force, see Fig. 4.4b. In this case, the parameter of interest is the drift velocity $v = \lim_{t \to \infty} \langle x(t) \rangle / t$.

4.3 X-Tilted Ratchet

We recall that in this case the correlated noise $f(t)$ is acting on the x variable. In Fig. 4.5 we present a schematic illustration of the rocking axis for the corresponding energy landscape in coordinates $(y - x, x)$. The implied tilting biases the states A', B' during the first half of the period and the states A, B during the second half of the period.

It will be convenient to rewrite our equations in the dimensionless form. We use the following definitions of the non-dimensional variables: $\tilde{X}(\tilde{t}) \equiv (1/a)X(t = \tau^*\tilde{t})$, $\tilde{V}(\tilde{x}, \tilde{y}) \equiv (1/(k_m a^2))V(x = a\tilde{x}, y = a\tilde{y})$, $\tilde{f}(\tilde{t}) \equiv (1/(k_m a))f(t)$ and $\tilde{\xi}_i(\tilde{t}) \equiv \xi_i(t)/\sqrt{\tau^*}$ Here τ^* is the main time scale of the problem $\tau^* = \eta_y/k_m$. The distance a between two minima of the potential $V(y-x)$ introduces the characteristic length scale, while the natural energy scale is $k_m a^2$. The remaining nondimensional parameters are $\tilde{D} \equiv k_B \Theta/(k_m a^2)$ and $\alpha = \eta_y/\eta_x$. Below, for simplicity, we omit the tildas.

In the soft device the dimensionless system takes the form:

$$\begin{cases} dx/dt = -\alpha \left[\partial_x \Phi(x) + \partial_x V(y - x) - f(t) \right] + \sqrt{2\alpha D} \, \xi_x(t), \\ dy/dt = -\partial_y V(y - x) + f_{ext} + \sqrt{2D} \, \xi_y(t), \end{cases} \tag{4.6}$$

while in the hard device we obtain

$$\begin{cases} dx/dt = -\alpha \left[\partial_x \Phi(x) + \partial_x V(y - x) - f(t) \right] + \sqrt{2\alpha D} \, \xi_x(t), \\ dy/dt = -\partial_y V(y - x) - k_m(y - z) + \sqrt{2D} \, \xi_y(t). \end{cases} \tag{4.7}$$

Fig. 4.5 X-tilted ratchet model: the energy landscape with a fictive axis showing how the rocking is applied

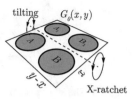

The Fokker–Planck equations corresponding to the normalized systems of Langevin equations (4.6) and (4.7) take the form:

- in the soft device configuration:

$$\partial_t P^s(x, y, t) = \alpha D \partial_x \left[\frac{P^s(x, y, t)}{D} \partial_x G^s(x, y, t) + \partial_x P^s(x, y, t) \right]$$
$$+ D \partial_y \left[\frac{P^s(x, y, t)}{D} \partial_y G^s(x, y, t) + \partial_y P^s(x, y, t) \right], \qquad (4.8)$$

where $G^s(x, y, t) = \Phi(x) + V(y - x) - x f(t) - y f_{ext}$.

- in the hard device configuration:

$$\partial_t P^h(x, y, t) = \alpha D \partial_x \left[\frac{P^h(x, y, t)}{D} \partial_x G^h(x, y, t) + \partial_x P^h(x, y, t) \right]$$
$$+ D \partial_y \left[\frac{P^h(x, y, t)}{D} \partial_y G^h(x, y, t) + \partial_y P^h(x, y, t) \right], \qquad (4.9)$$

where $G^h(x, y, t) = \Phi(x) + V(y - x) - x f(t) + \frac{1}{2} k_m (y - z)^2$.

In what follows, we adopt the simplest descriptions of the functions $\Phi(x)$, $V(y - x)$ and $f(t)$. Thus, we assume that

$$\Phi(x) = \begin{cases} \dfrac{\Phi_{max}}{\lambda_1}(x - nL), & \text{for } nL \leq x < nL + \lambda_1 \\[2ex] \dfrac{\Phi_{max}}{\lambda_2}((n+1)L - x), & \text{for } nL + \lambda_1 \leq x < (n+1)L, \ n \in \mathbb{N} \end{cases}$$
$$(4.10)$$

where $\Delta = \lambda_1 - \lambda_2$ is the parameter, which controls the potential asymmetry, when $\lambda_1 = (L + \Delta)/2$ and $\lambda_2 = (L - \Delta)/2$. The bi-stable element will be described by a piece-wise quadratic function

$$V(y - x) = \begin{cases} \frac{1}{2} k_0 (y - x)^2 + \varepsilon_0, & (y - x) \leq l, \\[2ex] \frac{1}{2} k_1 (y - x - a)^2, & (y - x) \geq l, \end{cases}$$
$$(4.11)$$

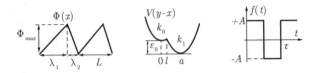

Fig. 4.6 The adopted simplified structure of the functions $\Phi(x), V(y-x)$ and $f(t)$

where $\varepsilon_0 = 1/2 \left(k_1(l-a)^2 - k_0 l^2\right)$. A periodic force $f(t)$ with period τ is assumed to be a square signal

$$f(t) = A(-1)^{n(t)} \quad \text{with} \quad n(t) = \lfloor 2t/\tau \rfloor, \tag{4.12}$$

where brackets $\lfloor \rfloor$ denote integer part. We illustrate the functions $\Phi(x)$, $V(y-x)$ and $f(t)$ in Fig. 4.6.

4.3.1 Typical Cycles

Soft Device We begin with a series of numerical experiments in the soft device. We perform computations by using the standard Euler scheme with the time step $\Delta t = 10^{-3}$ and then conduct the averaging over $N_r = 10^3$ realizations of the noise.

In Fig. 4.7a we illustrate the typical average trajectory of the X-tilted motor when the system reaches the steady state with the average velocity $v_x = v_y \equiv v$. Our Fig. 4.7b shows the time evolution of the system in coordinates $(t, y - x)$. Observe the emerging oscillations between the two wells of the bi-stable potential and note that the motor crosses a succession of space periods, see Fig. 4.7c.

We define the cycle as a segment of the average trajectory corresponding to one period of the driving $f(t)$. During the cycle we associate the transition $A'' \to B'$ with the release of the power stroke mechanism and the transition $B \to A''$ with its recharging.

In order to make sure that during each cycle the motor performs only one attachment-detachment step, we need to adjust our parameters. Suppose that $\alpha = 0.2$ (which controls the drift along x coordinate), take $L = 3$, and, in order to preserve the value of the force acting on the particle, choose $\Phi_{max} = 4.5$. In Fig. 4.8 we show the ensuing average behavior of the motor. As in the case shown in Fig. 4.7, the system reaches the steady state with a particular value of the average velocity, see Fig. 4.7a. The visible fluctuations can be explained by the relatively small number of stochastic realization in this case where we took $N_r = 200$.

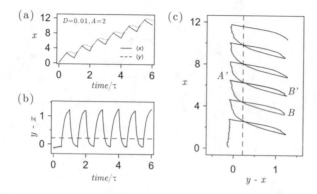

Fig. 4.7 The behavior of the X-tilted ratchet model in soft device with $f_{ext} = 0$. (a) The trajectories $\langle x(t)\rangle$ (solid black line) and $\langle y(t)\rangle$ (dashed gray line); (b) the time evolution of the system in coordinates $\{t, y - x\}$; (c) the average trajectory in coordinates $\{y - x, x\}$. Parameters: $k_0 = 1.5$, $k_1 = 0.43$, $l = 0.22$, $a = 1$, $\lambda_1 = 0.35$, $L = 0.5$, $\Phi_{max} = 0.75$, $\alpha = 1$, $A = 2$, $\tau = 16$

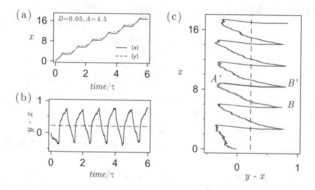

Fig. 4.8 The X-tilted ratchet in the soft device configuration with $f_{ext} = 0$. (a) The trajectories $\langle x(t)\rangle$ (solid black line) and $\langle y(t)\rangle$ (dashed gray line); (b) the time evolution of the system in coordinates $\{t, y - x\}$; (c) the average trajectory in coordinates $\{y - x, x\}$. The red line corresponds to $f(t) = +A$, blue line—to $f(t) = -A$. Parameters: $k_0 = 1.5$, $k_1 = 0.43$, $l = 0.22$, $a = 1$, $\lambda_1 = 2.4$, $L = 3$, $\alpha = 0.2$, $A = 4.5$, $\tau = 16$

The ensuing X motor cycle can be presented as a combination of two steps (see Fig. 4.9a):

- $1 \rightarrow 2$. First, because of the broken space symmetry, the motor advances in the x direction and crosses the energy barriers associated with the maxima of the periodic potential $\Phi(x)$. In the meanwhile the motor recharges the power stroke mechanism.

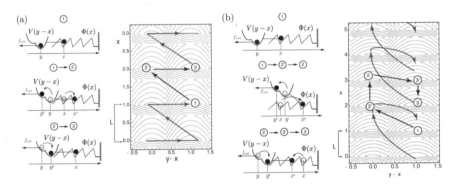

Fig. 4.9 The X-tilted ratchet model in the soft device configuration. With the solid gray lines we show the surface contours of the energy landscape $\Phi(x) + V(y-x)$. (**a**) The simplest motor cycle. (**b**) More complex motor cycle

- $2 \to 3$. As the rocking force changes its sign, the new configuration of the energy landscape drives the motor into the backward direction along the x axis. However, the motor is now trapped and instead of going backwards, it performs the power-stroke. Then the motor cycle starts again.

Note that in both studied cases the detachment and re-attachment take place simultaneously with the power stroke. Since the power stroke in this model is passive, the advance of the motor is due exclusively to the asymmetry of the potential $V(y-x)$: the motor must advance along the x axis in order to ensure the recharging of the power stroke mechanism. Note also that the ratchet must make a sufficiently large step in the forward direction in order to recharge the power stroke mechanism.

By varying the parameters of the model we can obtain other cycles as well. For instance, the motor step $1 \to 2$ can be made longer than a simple jump between the two nearest periods. Also, we can force the motor to move according to the scheme $1 \to 2' \to 2$, shown in Fig. 4.9b. Moreover, the motor can advance few periods along the x axis in the forward direction before accomplishing the power-stroke $2 \to 3$ and can also move backward during a few periods following the path $2 \to 3' \to 3$. Note, however that while both motor positions, $3'$ and 3, correspond to the same energy well 0 in the bi-stable potential, they are associated with different wells of the periodic potential.

Hard Device To study the effect of temperature we vary the parameter D while keeping the amplitude of the nonequilibrium driving fixed at $A = 6$ and in Fig. 4.10 we show the average trajectories during one time period of the force $f(t)$. With solid lines we show the energy landscape in the positive phase and with dashed black lines in the negative phase of rocking. We illustrate the average trajectory using different colors depending on the phase of rocking.

We observe that the cycle get stabilized after a short transient period. At low temperatures the response is localized around one minimum of the bi-stable

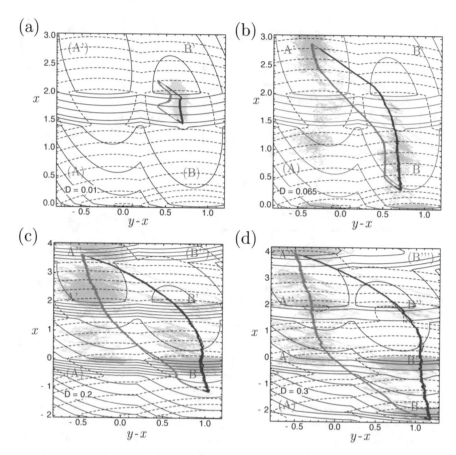

Fig. 4.10 The X-tilted ratchet model in the hard device configuration at different values of the temperature D and constant amplitude $A = 6$. The light gray lines depict a single stochastic realization during one time period. (**a**) $D = 0.01$; (**b**) $D = 0.065$; (**c**) $D = 0.08$; (**d**) $D = 0.3$. The other parameters are: $k_0 = k_1 = 7$, $l = 0.22$, $a = 1$, $L = 2$, $\lambda_1 = 1.4$, $\Phi_{max} = 5$, $\alpha = 5$ and $\tau = 20$

potential, see Fig. 4.10a. The increase of temperature allows the configurational particle to cross the potential barriers along both $y - x$ and x directions, in particular it allows for a transition between the energy wells of the bi-stable potential, see Fig. 4.10b, c. An (almost) three-state cycle is formed if the motor can displace itself along x sufficiently far in order to be able to recharge the power stroke element. If we increase the parameter D further, the particle makes larger jumps along the x direction, see Fog. 4.10d. At even higher temperatures the system loses its ability to generate force.

In Fig. 4.11 we present a schematic illustration of the attainable three-state cycle in the hard device. After a transient stage the motor follows the following trajectory in the clockwise direction $1 \rightarrow 2 \rightarrow 3 \rightarrow 1$ where:

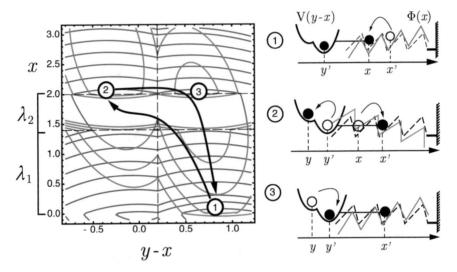

Fig. 4.11 The schematic representation of the typical X-tilted ratchet cycle in the hard device loading configuration. The average trajectory is shown by solid black arrows

- Transition $1 \rightarrow 2$. The average trajectory is shown schematically by black arrows. During the positive phase of the driving $f(t)$, motor crosses the barrier in the forward direction along the saw-tooth actin potential. It recharges the power stroke element, see Fig. 4.10 (phase plotted in red).
- During the negative phase of the driving $f(t)$ the motor makes the transition $2 \rightarrow 3 \rightarrow 1$, see Fig. 4.10 (phase plotted in blue). The change of sign of the force $f(t)$ biases the transition in the backward direction along the potential $\Phi(x)$. At the same time, the transition $2 \rightarrow 3$ takes place which we can identify with the power stroke. After the transition $3 \rightarrow 1$ the motor cycle starts again.

The main characteristic feature of the X model is that the power stroke is recharged purely mechanically, together with the advance of the motor along the actin filament. This unavoidably combines two different steps of the Lymn–Taylor cycle into one.

4.3.2 Force-Velocity Relations and Stochastic Energetics

Next we study the force-velocity relation for the X-tilted ratchet placed in the soft device. In Fig. 4.12 we present our numerical results for different temperatures D and two amplitudes of the non-equilibrium driving: $A = 2.5$ and $A = 4.5$. At $A = 2.5$ and small temperature we observe the characteristic concave shape of the force velocity curve. The area limited by the curve and the axes increases with temperature until the maximum is reached at $D = 0.1$. The subsequent increase

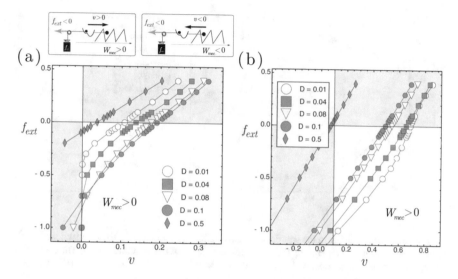

Fig. 4.12 Force-velocity relations for the X-tilted ratchet: (**a**) $A = 2.5$; (**b**) $A = 4.5$. The parameters: $k_0 = 1.5$, $k_1 = 0.43$, $l = 0.22$, $a = 1$, $L = 1$, $\lambda_1 = 0.7$, $\Phi_{max} = 1.5$, $\alpha = 1$, $\alpha = 1$ and $\tau = 20$

of temperature leads to the loss of the performance of the motor. The stall force, defined in soft device as the value of force at zero average velocity, increases with increasing temperature until the threshold value of D and decrease afterwards. At $A = 4.5$ we observe that the force velocity curves become convex. With the increase of temperature D we first approach the linear regime but then start to lose the active performance.

We define the active work performed by the motor against the load as $-f_{ext}v$ which means that it is positive when the average velocity and the external force have opposite signs. In these regimes the system is anti-dissipative, the motor uses (instead of dissipating) the external energy and can therefore perform some useful work, see the gray quadrants in 4.12. The predominantly passive regimes correspond to the cases where the average velocity and the external load have the same sign. Such systems are mostly dissipative and the energy is released rather than being absorbed, see the white quadrants in Fig. 4.12.

In Fig. 4.13a we present the average velocity of the motor as a function of temperature D for several values of the rocking amplitude A. At small amplitudes the motor shows a maximum of the velocity at a finite temperature. At high amplitudes the average velocity is a monotonically decreasing function of D which suggests that we deal with a purely mechanical ratchet. By light green color we identify the region with negative velocity where the motor is dragged by the cargo.

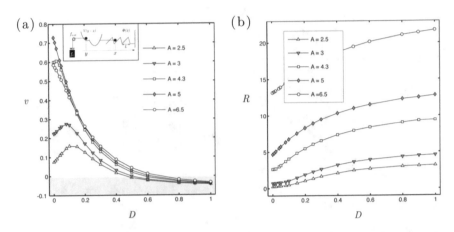

Fig. 4.13 X-tilted ratchet in the soft device configuration, $f_{ext} = -0.1$. (**a**) The dependence of the average velocity v on temperature D for the different values of the amplitude A. (**b**) The consumed energy R. The parameters: $k_0 = 1.5$, $k_1 = 0.43$, $l = 0.22$, $a = 1$, $L = 1$, $\lambda_1 = 0.7$, $\Phi_{max} = 1.5$, $\alpha = 1$ and $\tau = 30$

To study the energetics of the X-tilted ratchet we rewrite the equations of the model in the soft device (4.6) in the form

$$
\begin{cases}
\partial_x[\Phi(x) + V(y - x)] - f(t) = -\dfrac{1}{\alpha}d_t x + \sqrt{\dfrac{2D}{\alpha}}\,\xi_x(t), \\[2mm]
\partial_y V(y - x) - f_{ext} = -d_t y + \sqrt{2D}\,\xi_y(t).
\end{cases}
\tag{4.13}
$$

If we now multiply these equations by the vector $d\mathbf{X}$ in the Stratonovich sense and use the conventional definition of the exchanged heat $\delta Q_i = \left(-\eta d_t x_i + \sqrt{2\eta_i D}\xi_i(t)\right) \circ dx_i$ [105] we obtain

$$
\begin{cases}
\partial_x[\Phi(x) + V(y - x)] \circ dx - f(t) \circ dx = \delta Q_x, \\[2mm]
\partial_y V(y - x) \circ dy - f_{ext} \circ dy = \delta Q_y.
\end{cases}
\tag{4.14}
$$

Using the definition $G_0(x_t, y_t) = \Phi(x_t) + V(y_t - x_t) - y f_{ext}$ we can rewrite these equations in the form

$$
\begin{cases}
\partial_x G_0 \circ dx - f(t) \circ dx = \delta Q_x, \\[2mm]
\partial_y G_0 \circ dy = \delta Q_y.
\end{cases}
\tag{4.15}
$$

If we now average these equations over one time period τ, we obtain

$$
\frac{1}{\tau}\int_{\mathbf{X}_t}^{\mathbf{X}_{t+\tau}} dG_0(\mathbf{X}_t) - \frac{1}{\tau}\int_{x_t}^{x_{t+\tau}} f(t)dx_t = \frac{1}{\tau}\int_{\mathbf{X}_t}^{\mathbf{X}_{t+\tau}} (\delta Q_x + \delta Q_y).
\tag{4.16}
$$

Here we can identify the input energy $R = \frac{1}{\tau} \int_{x_t}^{x_{t+\tau}} f(t) dx_t$. Our Fig. 4.13b shows the value of R for the system in a soft device as a function of D for several values of A. One can see that with the increase of the level of thermal fluctuations the motor needs more energy in order to preform the work. At large temperatures we observe saturation, showing that the motor dragged by the cargo consumes energy at a fixed rate.

If we now define the (active or passive) mechanical work

$$W_{mec} = \frac{1}{\tau} \int_{\mathbf{X}_t}^{\mathbf{X}_{t+\tau}} dG_0(\mathbf{X}_t) = -\frac{1}{\tau} \int_{y_t}^{y_{t+\tau}} f_{ext} dy = -f_{ext} v_y \equiv -f_{ext} v, \tag{4.17}$$

we can then write the energy balance in the form $W_{mec} = R + Q$ where the heat term is $Q = \frac{1}{\tau} \int_{\mathbf{X}_t}^{\mathbf{X}_{t+\tau}} (\delta Q_x + \delta Q_y)$ [105]. For the system in the soft device, the mechanical efficiency of the system can be defined by the expression

$$\epsilon_{mec} = \frac{W_{mec}}{R}. \tag{4.18}$$

If we associate the functional work of an unloaded motor with overcoming viscous drag $W_{Stokes} = \alpha^{-1} v_x^2 + v_y^2$. we can also define the Stokes efficiency

$$\epsilon_{Stokes} = \frac{W_{Stokes}}{R}. \tag{4.19}$$

Finally, we can define the rectifying efficiency by combining Stokes and mechanical efficiencies:

$$\epsilon_{rec} = \frac{W_{mec} + W_{Stokes}}{R}. \tag{4.20}$$

In Fig. 4.14a we illustrate for the system in the soft device the mechanical work as a function of D at increasing values of A. By color, we mark the region of positive and negative mechanical work. In Fig. 4.14b we show the temperature dependence of the mechanical efficiency. In the regime of small amplitude A we observe a maximum of efficiency at finite temperature. With increasing amplitude A, the maximum vanishes and the efficiency becomes a monotonically decreasing function of D, which is the behavior characteristic for mechanical ratchets. By light green we indicate the regime of negative efficiency, where our motor is unable to perform a positive mechanic work against the external force and performs instead as an active breaking mechanism. In Fig. 4.14c we plot the Stokes efficiency as a function of D; the rectifying efficiency is shown in Fig. 4.14d. The shape of these functions is dominated by the quadratic Stokes term, however, the cumulative efficiency can

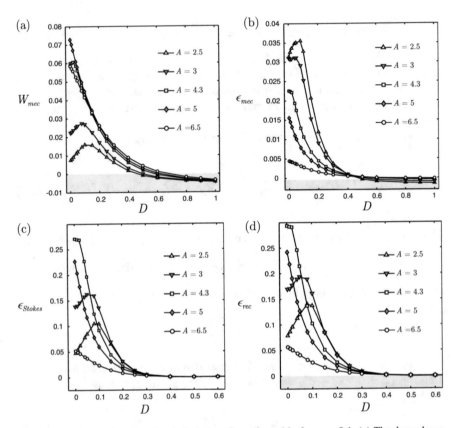

Fig. 4.14 X-tilted ratchet in the soft device configuration with $f_{ext} = -0.1$. (**a**) The dependence of the mechanical work W_{mec} on temperature D for different amplitudes A; (**b**) the mechanical efficiency ϵ_{mec}; (**c**) the stokes efficiency; (**d**) the rectifying efficiency. Parameters are the same as in Fig. 4.13

have a maximum at a finite temperature also if the amplitude of the rocking is small and the device works as a Brownian ratchet.

4.4 Y-Tilted Ratchet

The Y-tilted ratchet coupes the bi-stable potential $V(y - x)$ with the space periodic potential $\Phi(x)$. On the scheme presented in Fig. 4.15 we show again one period of the two-dimensional energy landscape with four mechanical configurations A, B, A', B' representing local minima of the energy. The applied periodic tilting acts along the diagonal and biases either the state B', during the positive phase of rocking, or the state A, during the negative phase.

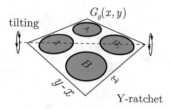

Fig. 4.15 Y-tilted ratchet energy landscape

The dimensionless system of Langevin equation for the Y-tilted motor in the soft device takes the form

$$
\begin{cases}
dx/dt = -\alpha\left[\partial_x \Phi(x) + \partial_x V(y - x)\right] + \sqrt{2\alpha D}\ \xi_x(t), \\
dy/dt = -\partial_y V(y - x) + f_{ext} + f(t) + \sqrt{2D}\ \xi_y(t).
\end{cases}
\tag{4.21}
$$

In the hard device we obtain

$$
\begin{cases}
dx/dt = -\alpha\left[\partial_x \Phi(x) + \partial_x V(y - x)\right] + \sqrt{2\alpha D}\ \xi_x(t), \\
dy/dt = -\partial_y V(y - x) - k_m(y - z) + f(t) + \sqrt{2D}\ \xi_y(t).
\end{cases}
\tag{4.22}
$$

Here we use the same potentials as before: $\Phi(x)$, see (4.10), $V(y - x)$, see (4.11) and $f(t)$, see (4.12). They are all schematically illustrated in Fig. 4.6.

4.4.1 Typical Cycles

Soft Device In Fig. 4.16 we show the two-dimensional representation of the motor trajectory in the case of zero external load, $f_{ext} = 0$. The main new observation is that the Y-tilted ratchet is able to generate a four-state cycle $A \to B \to B' \to A'$. This cycle is realistic and can be in principle directly compared with the biochemical Lymn–Taylor cycle.

The schematic representation of such cycle is shown in Fig. 4.17. It can be represented as a sequence of the following steps:

- $1 \to 1' \to 2$. We start at the end of the negative phase of the driving $f(t)$ when the system is in the state A. As $f(t)$ changes the phase (to positive), the energy switches to $\Phi(x) + V(y - x) - yA$. After an immediate advance $1 \to 1'$, the bi-stable element goes through the major transition $1' \to 2$ which we identify with the power stroke.
- $2 \to 3$. While the system remains in the positive phase of the loading $f(t)$, the motor makes a step along x direction from the state B to the state B'. This advance along the actin filament is the direct consequence of the power stroke which is here the driving force behind the detachment and reattachment.

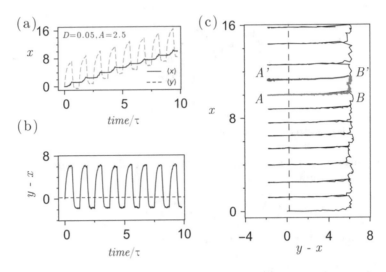

Fig. 4.16 Y-tilted ratchet in the soft device configuration with $f_{ext} = 0$. (**a**) The average trajectories $/\langle x(t) \rangle$ (solid black line) and $\langle y(t) \rangle$ (dashed gray line). (**b**) The time evolution of the system in coordinates $\{t, y - x\}$; (**c**) the average trajectory in coordinates $\{y - x, x\}$. The red lines correspond to the positive phase of the rocking $f(t) = +A$, the blue lines—to the negative phase of rocking $f = -A$. The parameters: $k_0 = 1.5$, $k_1 = 0.43$, $l = 0.35$, $a = 1$, $L = 1$, $\lambda_1 = 0.7$, $\Phi_{max} = 1.5$, $\alpha = 1$, $A = 2.5$, $\tau = 16$, $D = 0.06$

- $3 \rightarrow 3' \rightarrow 4$. Now the system is in state B', see Fig. 4.16d. The correlated force changes its sign and the energy landscape becomes tilted in the other direction. Following an immediate transition $3 \rightarrow 3'$ the power stroke element is recharged through the transition $3' \rightarrow 4$. Because of the asymmetry of the actin potential $\Phi(x)$ the coordinate x gets trapped and does not move in the backward direction. Therefore the advance along the actin filament has taken place and the cycle can start again.

Depending on the amplitude of the correlated noise term, the motor step $2 \rightarrow 3$ can be longer or shorter. In particular, the system can jump over several periods of the potential $\Phi(x)$. The length of such 'step' is influenced by the fine structure of the energy landscape and also depends on the stiffness of the bi-stable spring.

Hard Device We assume that $z = 0$. The cycles emerging after short transients are illustrated in Fig. 4.18.

At small temperature $D = 0.01$ we observe oscillations between the conformational states A and B inside the same space period of the actin potential ($x = 0$), see Fig. 4.18a. This behavior can be interpreted as a power-stroke (red path) followed by the recharging (blue path) in the attached state. The little loop around the state B is a consequence of the distorted landscape in the hard device. We can therefore speak here about a two-state cycle. With the increase of temperature the Brownian particle is able to explore larger areas of the two-dimensional landscape and at $D = 0.1$ we can stabilize the oscillations between the state A' and B', see Fig. 4.18b. In this case

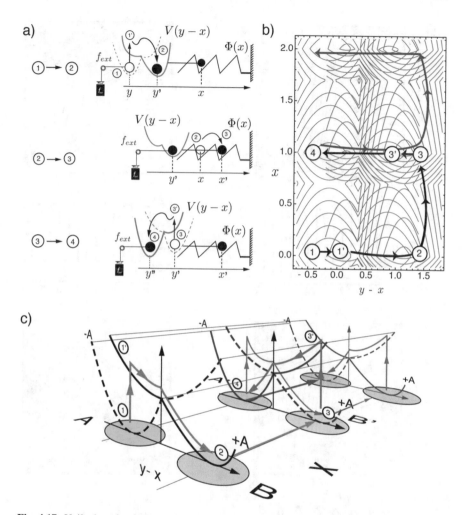

Fig. 4.17 Y-tilted ratchet in the soft device configuration. (**a**) Mechanical representation of the motor cycle. (**b**) Average trajectory superposed with the energy landscape in the positive (solid red lines) and in the negative (solid blue lines) phases of the loading. (**c**) Schematic representation of the major transitions associated with the power stroke element

the system reattaches to a new cite on the actin filament and stretches the spring. As a result, the motor generates much larger average tension, however, the cycle is still composed of only two states. By increasing the temperature further, we are forcing the system to go through a four state cycle, see Fig. 4.18c. Interestingly, in this regime the motor generates smaller tension then in the regime of slightly lower temperature when the cycle consists of two states only. At even higher values of D we still have the four-state cycle but we now occasionally encounter also disadvantageous transitions in the backward direction, see Fig. 4.18d.

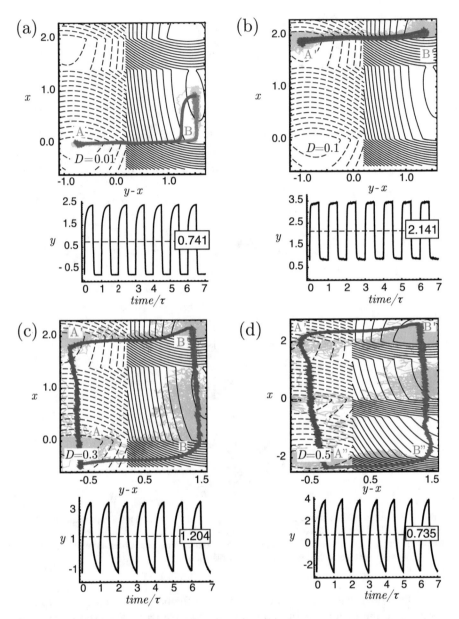

Fig. 4.18 Y-tilted ratchet in the hard device configuration at different temperatures and constant amplitude of non-equilibrium driving $A = 6$. Solid lines represent the energy landscape in the positive and dashed lines—in the negative phase of rocking. The average motor trajectory is shown by the thick red line during the positive and by the thick blue line during the negative phase. The light gray lines follow the single stochastic realization during one time period: (**a**) $D = 0.01$; (**b**) $D = 0.1$; (**c**) $D = 0.3$; (**d**) $D = 0.8$. The parameters: $k_0 = k_1 = 7$, $l = 0.22$, $a = 1$, $L = 2$, $\lambda_1 = 1.4$, $\Phi_{max} = 5$, $\alpha = 5$, $k_m = 1$ and $\tau = 20$

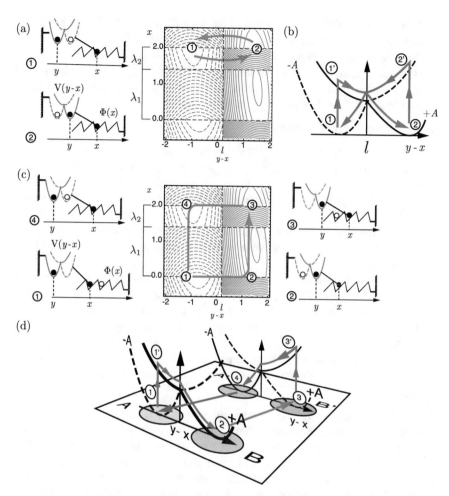

Fig. 4.19 Y-tilted ratchet in the hard device configuration. With solid gray lines we show the energy level contours in positive phases of the rocking when potential is $\Phi(x) + V(y - x) + 1/2k_m y^2 - yA$ while with dashed gray lines we show the energy level contours in the negative phase of rocking when the potential is $\Phi(x) + V(y-x) + 1/2k_m y^2 + yA$. (**a**) The two-state motor cycle, the average trajectory is shown schematically by red arrows in the positive phase and by blue arrows in the negative phase; (**b**) schematic structure of the two-state cycle; (**c**) the four-state motor cycle; (**d**) schematic structure of the four-states cycle

In Fig. 4.19 we present a more systematic comparison of the observed two-state and four-state cycles. In the 'two- state cycle' we observe the following steps:

- $1 \rightarrow 1' \rightarrow 2$. We start at the end of the phase with the negative tilt. The system explores the state A, see Fig. 4.19b. As the tilt switches to positive, the system makes an immediate transition $1 \rightarrow 1'$. From this new configuration the system performs the power stroke $1' \rightarrow 2$. The motor then remain in the state B.

- $2 \rightarrow 2' \rightarrow 1$. While the system is in the state B, see Fig. 4.19b, the correlated noise changes sign, creating again the negative tilt of the energy landscape. The system undergoes an immediate transition $2 \rightarrow 2'$. Because of the spatial asymmetry of the actin potential the system remains trapped in the same period of the periodic potential $\Phi(x)$ while the power stroke is recharged performing the transition $2' \rightarrow 1$. Then the cycle starts again.

In the 'four-state cycle' the steps are:

- $1 \rightarrow 1' \rightarrow$. We start again at the end of the negative phase of the rocking. The system explores the state A. As the noise term $f(t)$ changes the value from negative to positive, the system makes an immediate transition $1 \rightarrow 1'$, see Fig. 4.19d. From this new configuration the system performs the power stroke $1' \rightarrow 2$.
- $2 \rightarrow 3$. While in the positive phase of $f(t)$, the system makes a jump into the next nearest well in the positive x direction which is the consequence of the power stroke.
- $3 \rightarrow 3' \rightarrow 4$. The system is now in the state 3 and the corresponding energy well is B', see Fig. 4.19d. The correlated noise term changes the sign and the system undergoes a transition $3 \rightarrow 3'$. Then the instability causes the particle to perform the transition $3' \rightarrow 4$, which we interpret as the recharging of the power stroke mechanism.
- $4 \rightarrow 1$. From the state A' the motor jumps in backward direction making the transition $4 \rightarrow 1$. The system returns into the initial state and the cycle can start again.

To summarize, in the two-state regime the system is residing in a distant, force generating well of the periodic potential while performing periodic oscillations between the two conformational states of the power stroke element. The level of the generate force is high because the cross bridge is firmly attached throughout the cycle. In the four-state regime, the system is periodically reaching the distant well of the periodic potential but remains there only for a limited time before returning back to the original attachment site. In this regime the average force is smaller, however the mechanical cycle is closer to its biochemical analog.

4.4.2 Force-Velocity Relations and Stochastic Energetics

In Fig. 4.20a we show the force-velocity relation at different temperatures D and fixed amplitude of rocking $A = 2.5$. Observe that it is mostly convex at this level of driving. With the increase of temperature the system generates smaller average velocity at zero load and is characterized by smaller stall force. These trends were similar in the case of X-tilted ratchet, see Fig. 4.12. In Fig. 4.20b we show how the force-velocity relation changes when we vary the amplitude of rocking A at the fixed temperature $D = 0.01$. We obtain concave force-velocity relations at small

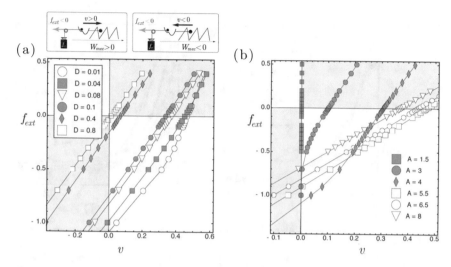

Fig. 4.20 Force velocity relations for the Y-tilted ratchet: (**a**) temperature dependence at $A = 2.5$; (**b**) non-equilibrium noise amplitude dependence at the temperature $D = 0.01$

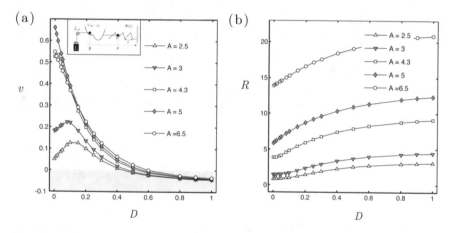

Fig. 4.21 Y-tilted ratchet in the soft device configuration working against the load $f_{ext} = -0.1$: (**a**) temperature dependence of the drift velocity at different values of A; (**b**) Consumed energy R dependence on temperature D at different values of A. The parameters $k_0 = 1.5$, $k_1 = 0.43$, $l = 0.22$, $a = 1$, $L = 1$, $\lambda_1 = 0.7$, $\Phi_{max} = 1.5$, $\alpha = 1$ and $\tau = 60$

amplitude of the driving where we expect that the system to work as a thermal ratchet but then recover the convexity in the interval $1.5 < A \leq 3$.

In Fig. 4.21a we show the average velocity as a function of temperature at different values of the amplitude A. For small amplitudes (thermal ratchet regime) the motor exhibits a maximum of velocity at finite temperature. At higher amplitudes A (mechanical ratchet) the average velocity decreases monotonically with D. Overall the Y-tilted model is generating smaller average velocities than the X-tilted model at the same values of parameters.

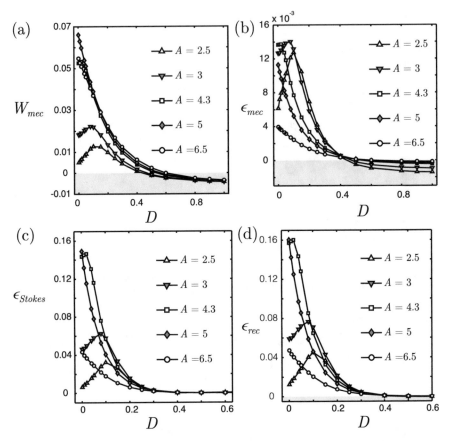

Fig. 4.22 Y-tilted ratchet in the soft device working against the load $f_{ext} = -0.1$. Temperature dependence at different values of A: (**a**) the mechanic work W_{mec}; (**b**) the mechanic efficiency; (**c**) the Stokes efficiency; (**d**) the rectifying efficiency. The parameters are the same as in Fig. 4.21

In Fig. 4.21b we show the consumed energy R as a function of D and use the same range of amplitudes A. As the temperature increases, the motor consumes more and more energy in order to preform useful work. We also again observe a saturation at high temperatures meaning that there is a limit of how much thermal energy the motor can rectify.

Finally, in Fig. 4.22 we show the temperature dependence of the various measures of efficiency at increasing values of A. The qualitative behavior of all these functions is basically the same as in the case of X-tilted ratchet.

Fig. 4.23 XY-tilted ratchet:
the effective tilting of the
energy landscape

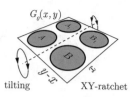

4.5 XY-Tilted Ratchet

In the XY-tilted ratchet the correlated force $f(t)$ acts on the combination of variables
$y - x$, which can be identified with an internal strain inside the bi-stable element.
This means that the rocking force affects the power stroke mechanism directly
instead of implicitly modifying the internal state of this device through other
external degrees of freedom. In Fig. 4.23 we illustrate the mechanical action of the
rocking force on the two dimensional energy landscape: note that the ATP activity
is now fully decoupled from the actin filament.

In the non-dimensional variables the main system of equations describing the
activity of the XY-tilted ratchet in the case of soft device takes the form

$$
\begin{cases}
dx/dt = -\alpha\,[\partial_x\Phi(x) + \partial_x V(y - x) + f(t)] + \sqrt{2\alpha D}\,\xi_x(t), \\
dy/dt = -\partial_y V(y - x) + f_{\text{ext}} + f(t) + \sqrt{2D}\,\xi_y(t).
\end{cases}
\tag{4.23}
$$

In the case of hard device we obtain

$$
\begin{cases}
dx/dt = -\alpha\,[\partial_x\Phi(x) + \partial_x V(y - x) + f(t)] + \sqrt{2\alpha D}\,\xi_x(t), \\
dy/dt = -\partial_y V(y - x) - k_m(y - z) + f(t) + \sqrt{2D}\,\xi_y(t).
\end{cases}
\tag{4.24}
$$

In the corresponding Fokker–Planck equations (4.8) and (4.9) we must use the
potential $G^s(x, y, t) = \Phi(x) + V(y - x) - (y - x)f(t) - yf_{\text{ext}}$ in the case of
the soft device and $G^h(x, y, t) = \Phi(x) + V(y - x) - (y - x)f(t) + \frac{1}{2}k_m(y - z)^2$ in
the case of the hard device. In our numerical experiments we used the same choices
for the functions $\Phi(x)$, see (4.10), $V(y - x)$, see (4.11) and $f(t)$, see (4.12) as in
the previous sections, see Fig. 4.6.

Before turning to the structure of the generated cycles we remark that a
conceptually similar approach was used before to describe Kinesin modeled as two
coupled rocking ratchets which move along the same periodic potential [78]. The
corresponding system of over-damped Langevin equation can be written ass:

$$
\begin{cases}
dx/dt = -\partial_x\Phi(x) - \partial_x V(x - y) - f(t) + \sqrt{2D}\,\xi_x(t), \\
dy/dt = -\partial_y\Phi(y) - \partial_y V(x - y) + f(t) + \sqrt{2D}\,\xi_y(t),
\end{cases}
\tag{4.25}
$$

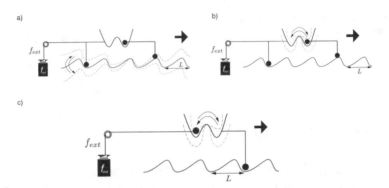

Fig. 4.24 Kinesin-type molecular motors: (**a**) ratchet concept developed in [79]; (**b**) ratchet concept developed in [78]; (**c**) ratchet concept developed here, XY-tilted ratchet

where $\Phi(x)$ and $\Phi(y)$ are two identical ratchet potentials, $V(x - y)$ is the bistable potential describing the interaction between the two legs of the Kinesin motor, whose positions are given by coordinates x and y. Note that here, as in our model, the time periodic rocking acts on the coordinate $x - y$, which indicates the implicit activity of the bistable element, see Fig. 4.24. In contrast to such 'two-leg' designs describing processive motors, our 'one-leg' design concerns non-processive motors.

4.5.1 Motor Cycles

Soft Device In Fig. 4.25 we show the averaged trajectory of the XY-tilted ratchet exposed to a rocking force with amplitude $A = 0.6$ and simultaneously subjected to a thermal noise with $D = 0.02$. The system follows a three-state cycle: it performs a power-stroke while being attached to one particular state on the periodic landscape and then moves in the forward direction along this landscape, while recharging the power-stroke mechanism. The change of sign of the tilting force $f(t)$ both re-activates the power-stroke mechanism and causes the directional motion of the motor. The amplitude of the tilting strongly influences the shape of the energy landscape, in particular, in each phase, positive and negative, the intrinsic bi-stability of the potential in the $y - x$ direction may be either preserved or not. In Fig. 4.26 we schematically show the states visited by the motor during one cycle:

- $1 \rightarrow 1' \rightarrow 2$. We start the cycle at the very end of the negative phase of the rocking, see Fig. 4.26 when the system explores the state A. As the force $f(t)$ changes from negative to positive, the energy becomes $\Phi(x)+V(y-x)-(y-x)A$ and the particle makes a transition $1 \rightarrow 1'$. During the positive phase of the rocking the system undergoes the transition $1' \rightarrow 2$ which we identify with the power stroke.

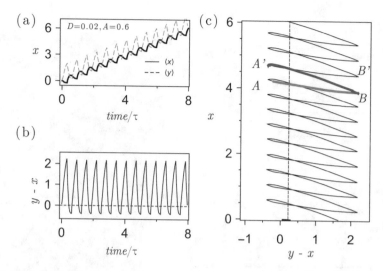

Fig. 4.25 XY-tilted ratchet in the soft device configuration with $f_{ext} = 0$: (**a**) Trajectories $\langle x(t) \rangle$ (solid black line) and $\langle y(t) \rangle$ (dashed gray line); (**b**) time evolution of the system in coordinates $\{t, y - x\}$; (**c**) average trajectory in coordinates $\{y - x, x\}$; red line correspond to the rocking phase $f(t) = +A$, blue lines—to the phase $f = -A$. The parameters $k_0 = 1.5$, $k_1 = 0.43$, $l = 0.35$, $a = 1$, $\lambda_1 = 0.35$, $L = 0.5$, $\Phi_{max} = 1$, $\alpha = 1$ and $\tau = 16$

- $2 \to 3' \to 3$. We are now in state 2 corresponding to the state B, see Fig. 4.25. The rocking force changes the sign and the energy becomes $\Phi(x) + V(y - x) + (y - x)A$. The system makes the step $2 \to 3'$, and since the bistable potential is now biased, the motor performs an additional transition $3' \to 3$, finalizing the recharging of the power stroke element. Because of the spatial asymmetry of the periodic potential the attachment site does not change during such recharging. Then the cycle starts again.

Note that the motor advance and the recharging of the power stroke take place simultaneously. Those are the two stages where the external energy supply is necessary and they cannot be separated in this setting.

Hard Device We fix the total displacement at $z = 0$ and show in Fig. 4.27 the simplest motor cycles. At low temperatures the system moves only between the states A and B. This means that the cross bridge is attached to a particular site of the actin potential while performing random work between two configurations of the power stroke element. With the increase of temperature D the motor eventually crosses the energy barrier (detaches) and then stabilizes (attaches) in the next site on the actin filament. In this new attached position the motor continues to perform the transitions between state A' and state B', see Fig. 4.27b. Observe that now the attachment site is distant from the reference position, the spring is stretched and the motors shows higher levels of tension comparing to the cycle shown in Fig. 4.27a.

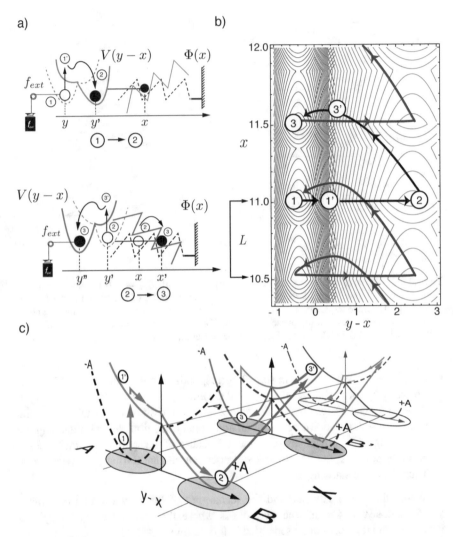

Fig. 4.26 Schematic cycle of the XY-tilted ratchet in the soft device configuration: (**a**) mechanical representation of the motor cycle; (**b**) averaged trajectory; (**c**) the cycle showing the energy changes associated with different moves of the system in the space $((y - x), x)$

By increasing the temperature and the amplitude of rocking further, we force the motor to visit more sites on the energy landscape. Thus in Fig. 4.27c the motor periodically changes the attachment site: the cycle is performed between the state A' and the state B, corresponding to different attachment sites along the potential $\Phi(x)$. In this regime the motor is able to generate the highest level of average tension. We see first motor detachment and advance, accompanied by the recharging of the power stroke element, and then the power stroke combined with re-attachment

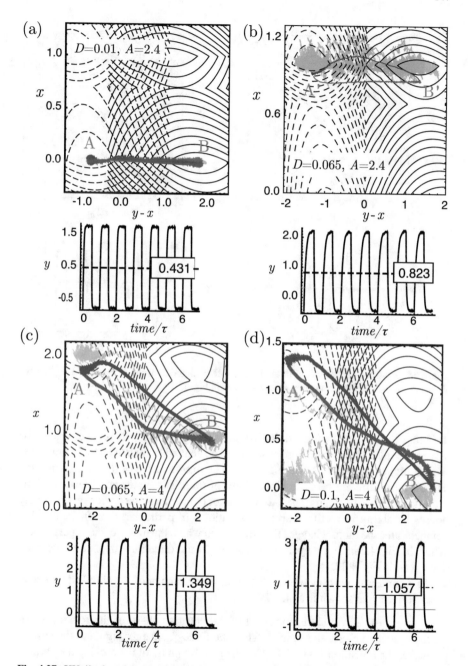

Fig. 4.27 XY-tilted ratchet model in the hard device configuration with $z = 0$. The dependence of the motor cycle on the temperature D and the rocking amplitude A: (**a**) $D = 0.01$, $A = 2.4$; (**b**) $D = 0.065$, $A = 2.4$; (**c**) $D = 0.08$, $A = 4$; (**d**) $D = 0.1$, $A = 4$. The parameters $k_0 = 1.5$, $k_1 = 0.43$, $l = 0.35$, $a = 1$, $\lambda_1 = 0.7$, $L = 1$, $\Phi_{max} = 1.5$, $\alpha = 1$ and $\tau = 10$

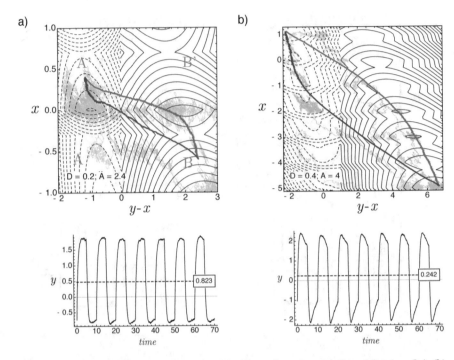

Fig. 4.28 XY-tilted ratchet model in the hard device configuration: (**a**) $D = 0.2$, $A = 2.4$; (**b**) $D = 0.4$; $A = 4$. Other parameters are the same as in Fig. 4.27

bringing the system to the original site. Once again, the two biochemical steps appear coupled in this mechanical setting. The slanted two-state cycle is preserved also at higher values of D, see Fig. 4.27d. The fine structure of the cycle, however, is now a bit different because higher level of noise stimulates additional spurious transitions inside the landscape. At even larger temperature the device progressively looses its ability to rectify thermal fluctuations. In particular, at sufficiently high temperatures the motor changes the direction in which the motor cycle is performed. Thus, in Fig. 4.28a, b the cycle is performed in the direction opposite to what we have seen in Fig. 4.27.

To summarize, we now present the schematic structure of the simplest two-state hard device cycle shown in Fig. 4.29.

- $1 \to 1' \to 2$. We start at the end of the negative phase of the rocking when the system explores the state A. Then the force f changes sign and the system makes the transition $1 \to 1'$. During the positive phase of the rocking the particle performs the transition $1' \to 2$, which we associate with the power stroke.
- $2 \to 2' \to 1$. We are now in the state B. The rocking force changes sign and the system makes the transition $2 \to 2'$. Because of the spatial asymmetry of the periodic potential, the system remains trapped in the 'distant minimum' of $\Phi(x)$

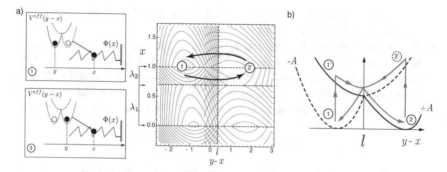

Fig. 4.29 Schematic representation of the XY-tilted ratchet cycle in the hard device corresponding to the trajectory shown in Fig. 4.27a: (**a**) two-state motor cycle; (**b**) energetics of the two-state cycle

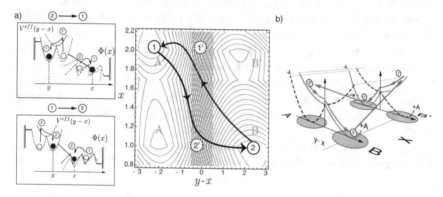

Fig. 4.30 Schematic representation of the XY-tilted ratchet cycle in the hard device corresponding to the trajectory shown in Fig. 4.27d: (**a**) 'slanted' two-state motor cycle; (**b**) energetics of the 'slanted' two-state cycle

while performing the transition $2' \to 1$, which we interpret as the recharging of the power stroke. Then the cycle can start again.

In Fig. 4.30 we similarly illustrate the (low temperature) 'slanted' two-state cycle corresponding to what we have seen in Fig. 4.27d. Formally, the motor visits only two sites corresponding to the stable states A' and B', see Fig. 4.30a. However, because of the peculiar shape of the cycle we can distinguish two additional intermediate states marked as $1'$ and $2'$. With these additional states taken into consideration we can interpret the ensuing periodic trajectory as the following four-states cycle:

- $1 \to 2' \to 2$. We start again at the end of the negative phase of the rocking when the system explores the state A'. Then the force f changes its sign from negative to positive and the system makes a transition $1 \to 2'$ 'moving back' along the x coordinate. During the positive phase of the rocking the system performs an additional transition $2' \rightarrow 2$ and finds itself in the state B. We interpret the transition $2' \rightarrow 2$ as the power stroke.

- $2 \rightarrow 1' \rightarrow 1$. We now are in the state 2, point B. The correlated force changes its sign again, from positive to negative. The system detaches and makes a 'forward jump' to the next site along the actin filament performing the transition $2 \rightarrow 1'$. Subsequently the particle continues with the transition $1' \rightarrow 1$ which we interpret as the recharging of the power stroke mechanism. Then, the cycle can start again.

4.5.2   Force-Velocity Relations and Stochastic Energetics

In Fig. 4.31a we show the effect of temperature at low amplitudes of rocking $A = 2.5$ on the force velocity relations for a XY-tilted ratchet. At zero temperature the system exhibits purely mechanical behavior without any 'anti-dissipation' (no entrance into the white quadrants). The growth of temperature D increases the area between the force-velocity curve and the axes in the domain of anti-dissipative behavior. We observe the pronounced concave character of the force-velocity relations at sufficiently low temperatures. After the threshold in D the concavity progressively vanishes and the profile becomes linear, while the motor loses its ability to carry the cargo. In Fig. 4.31b we illustrate the dependence of the force-velocity relation on A at fixed $D = 0.02$. At small amplitudes of rocking the motor follows closely the external force f_{ext} and does not perform useful mechanical work.

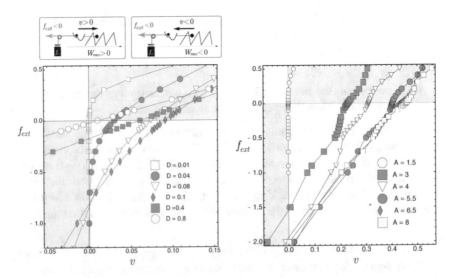

Fig. 4.31 The XY-tilted ratchet in soft device configuration: (**a**) dependence of the force-velocity relation on temperature D at the fixed $A = 2.5$; (**b**) dependence of the force-velocity relation on A at the fixed temperature $D = 0.02$. Parameters are the same as in Fig. 4.27

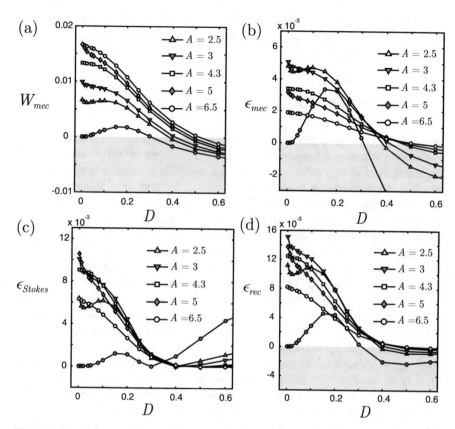

Fig. 4.32 XY-tilted ratchet in the soft device configuration working against the fixed load $f_{ext} = -0.1$. (**a**) the mechanical work W_{mec}; (**b**) the mechanical efficiency; (**c**) the Stokes efficiency; (**d**) the rectifying efficiency. Parameters are the same as in Fig. 4.27

Only after a certain threshold in amplitude the motor starts to generate active drift against the load.

In Fig. 4.32a we show the mechanical work as a function of D at different amplitudes of the rocking amplitude A. We observe two regimes: with positive and with negative mechanic work. In Fig. 4.32b we present the mechanical efficiency. At small amplitudes A we observe a maximum of the efficiency at finite temperature. With increasing A the maximum vanishes and the efficiency becomes a monotonically decreasing function of D, as one can expect in a purely mechanic ratchet. By light green color we indicated the regimes with the negative efficiency, where dissipation prevails. In Fig. 4.32c we present the Stokes efficiency as a function of D. We observe maxima on the efficiency vs D curve corresponding to finite temperatures and low rocking amplitudes regimes and also see that in some regimes the Stokes efficiency may increase with temperature. The rectifying efficiency is shown in Fig. 4.32d. Once again, at small amplitudes of rocking we see thermal ratchet behavior with a maximum of efficiency at finite temperature while at high rocking amplitudes we see the mechanical ratchet behavior with efficiency decreasing with temperature.

4.6 Comparison of the Three Models

We introduced above three minimal mechanistic models which all describe muscle contraction in terms of Langevin dynamics of a mechanical system. By localizing the mechanical effects of the ATP-related activity on a single internal degree of freedom, we could study separately the possibilities that actin filament is active (X model), that the coupling between the attachment/detachment and the power stroke elements is active (Y model) and that the power stroke element itself is active (XY model). Now we chose a single set of parameters and compare directly the resulting force-velocity relations and the efficiencies of the associated energy transduction mechanisms. To allow such a comparison we continue to use the simplest descriptions of the functions $\Phi(x)$, $V(y-x)$ and $f(t)$ shown in Fig. 4.6.

To produce a realistic description of muscle contraction we use the time scale $t^\star = \eta_y/k_m \sim 0.2$ ms where $\eta_y \sim 0.38$ ms $\cdot$ pN/nm is the micro-scale viscosity [19] and where $k_m \sim 2$ pN/nm is the stiffness of the elastic part of the myosin motor [7, 67]. The spatial scale is then $l^\star = a$ where $a \sim 10$ nm is the distance between two minima of the pre and post power stroke wells [69]. Then the energy scale is $\epsilon^\star = k_m a^2 \sim 200$ pN $\cdot$ nm. We assume that $D = k_B \Theta/(k_m a^2) \sim 0.02$ where $k_B = 4.10$ pN $\cdot$ nm is the Boltzmann constant, $\Theta \sim 300$ K is the ambient temperature, and $a = 10$ nm is the characteristic size of a motor power-stroke [69]. For the active driving we obtain $\tau = \tau_{ATP}/(\eta/k_m) \sim 100$ where $\tau_{ATP} = 40$ ms is the characteristic time of ATP hydrolysis [53]. We can now write $A = \sqrt{\Delta\mu/(k_m a^2)} \approx 1$ where $\Delta\mu = 20 k_B \Theta$ is the typical value of degree of non-equilibrium in terms of the affinity of ATP hydrolysis reaction [53]. Finally we assume that the non-dimensional parameters of the bi-stable potential take the values [19]: $k_0 = 1.5$, $k_1 = 0.43$, $l = 0.35$, $a = 1$. The space periodic potential is characterized by the parameters $\lambda_1 = 0.7$, $L = 1$, $\Phi_{max} = 1.5$ and in most illustrations we suppose for simplicity that $\alpha = 1$.

4.6.1 Soft Device

In Fig. 4.33 we compare the drift velocities generated by X, Y and XY-tilted ratchets in the soft device. The common feature of all three systems is that they exhibit the phenomenon of stochastic resonance: the average velocity is maximized at a particular value of temperature. At large temperatures all three systems progressively lose the capacity to rectify thermal fluctuations. In the regimes with high amplitude of the rocking the stochastic resonance disappears and the average

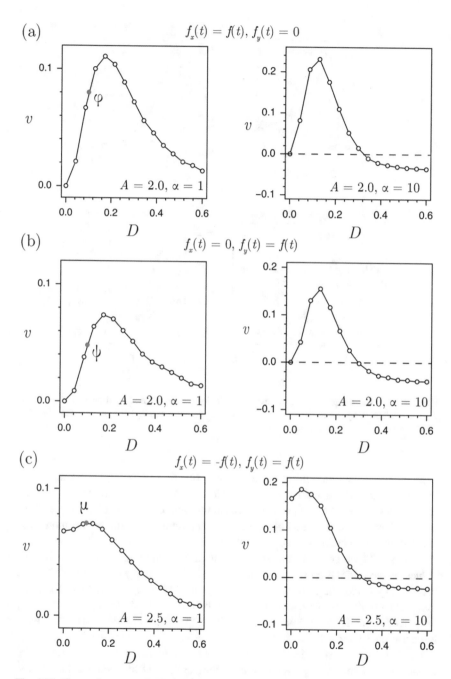

Fig. 4.33 The performance of different motors in soft device conditions with $f_{ext} = 0$. Average velocity v as a function of temperature D at different values of the amplitude A for X ratchet (**a**), Y ratchet (**b**), XY ratchet (**c**). Here the correlation time $\tau = 30$

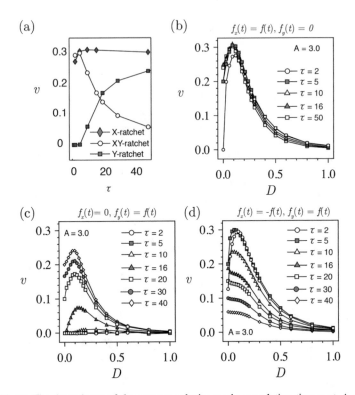

Fig. 4.34 (a) The dependence of the average velocity on the correlation time τ at $A = 3$ and $D = 0.08$. The variation of the average velocity with temperature D for X-ratchet (**b**), Y-ratchet (**c**) and XY-ratchet (**d**). Here $\alpha = 1$

velocity becomes a monotonically decreasing function of D. This is an indication that in all three cases the nature of the ratchet changes from Brownian to purely mechanical. Observe also that at $\alpha = 10$ (in the right column of Fig. 4.33) all three motors change their direction of the motion because of an additional asymmetry in the system which competes with polarity of the actin filament.

In Fig. 4.34 we compare the average velocities in the three models at different values of the correlation time τ characterizing the driving $f(t)$. In the case of X ratchet, Fig. 4.34a, the influence of τ is minimal. This suggests that for small and moderate amplitudes of rocking, the ratchet behavior can be well approximated by the effective model with adiababtically eliminated fast variable y. The Y ratchet is operational over a broad interval of the periods τ. Instead, the XY ratchet is functional only for sufficiently small values of τ. In Fig. 4.35, we compare the dependence of the average velocity v on temperature D at several values of the rocking amplitude A.

In Fig. 4.36 in the left column we compare the average trajectories for X, Y and XY ratchets at the points φ, ψ and μ marked in Fig. 4.33. As before, we use red color to identify the part of the cycle associated with the positive phase of the rocking, and

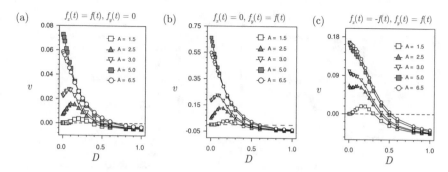

Fig. 4.35 The dependence of the average velocity v on temperature D at $f_{ext} = -0.1$; X ratchet (**a**); Y ratchet (**b**); XY ratchet (**c**). The parameters: $\alpha = 1$ and $\tau = 30$

blue color for the path associated with the negative phase. The dotted line shows the boundary between left and right wells of the bi-stable potential and we associate its crossing with either release or re-cocking of the power stroke mechanism.

For the X-tilted ratchet shown in Fig. 4.36a we obtain a three-state cycle, where the detachment and the re-attachment take place simultaneously with the recharge of the power stroke. For the Y-tilted ratchet shown in Fig. 4.36b we obtain a four -state cycle. Note that here, the motor releases and recharges the power-stroke mechanism actively, moreover without the power-stroke activity the motor won't be able to move. Finally, for the XY -tilted ratchet shown in Fig. 4.36c we again obtain a three-state cycle. First, it performs a power-stroke while being attached to one particular site on the periodic landscape $\Phi(x)$ and after that it moves forward along the potential $\Phi(x)$, while in same time recharging the power-stroke mechanism. Here again the motor advance and the recharging of the power stroke take place simultaneously.

In Fig. 4.37 we compare the force velocity relations at different temperatures D and fixed rocking amplitude $A = 2.5$. At zero temperature all three systems exhibit purely mechanical depinning behavior without showing any "anti-dissipation". At finite temperatures we obtain the concave force-velocity relations in agreement with experimental observations [62, 87, 125]. After the threshold $D \approx 0.1$ the force velocity relations becomes almost linear and eventually the motors lose their ability to carry cargo. The XY-tilted ratchet shows the highest stall force value among the three type of devices.

4.6.2 Hard Device

In Fig. 4.38 we compare the average tension generated by the motors in the hard device at different temperatures D and rocking amplitudes amplitude A. The active tension is optimized at a finite value of D for all configurations given that the amplitude of the correlated noise is sufficiently low.

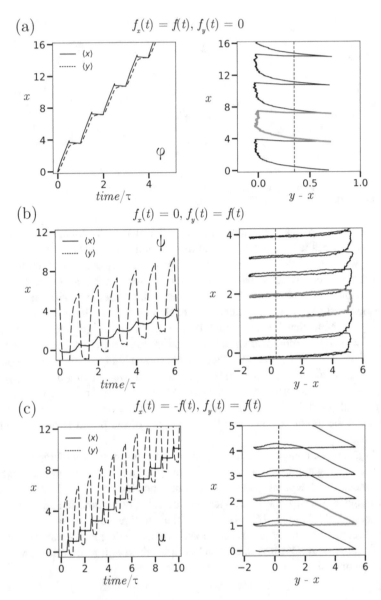

Fig. 4.36 Typical trajectories of the three motors in the soft device with $f_{ext} = 0$: (**a**) X-tilted ratchet; (**b**) Y-tilted ratchet; (**c**) XY-tilted ratchet. The parameters: $\alpha = 1$ and $\tau = 30$

At low temperatures and low amplitudes of rocking the X-tilted ratchet generates small tension because the energy transmitted to the motor is not sufficient to activate the bi-stable element. The Y-tilted ratchet shows the plateau regimes, where the system acts as simple mechanical bi-stable element. The XY-tilted

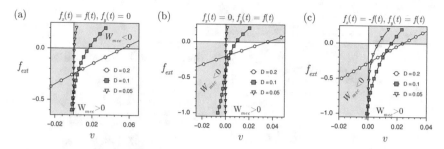

Fig. 4.37 Comparison of the force-velocity relations at fixed $A = 1.5$ for: X ratchet (**a**), Y ratchet (**b**), XY ratchet (**c**). The parameters: $\alpha = 1$ and $\tau = 30$

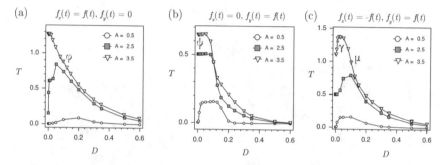

Fig. 4.38 The dependence of the average tension T on temperature D for: X-tilted ratchet (**a**), Y-tilted ratchet (**b**) and XY -tilted ratchet (**c**). The parameters $k_0 = 1.5$, $k_1 = 0.43$, $l = 0.35$, $L = 1, \lambda_1 = 0.7$, $\Phi_{max} = 1.5, \tau = 30, \alpha = 1, k_m = 1$ and $z = 0$. Under these conditions the averaged tension T is equal to $\langle\langle y \rangle\rangle$

ratchet demonstrates a hybrid behavior exhibiting the temperature and the amplitude thresholds whose crossing allows the motor to form a cycle. Such regime takes advantage of both, the thermal fluctuations and the correlated noise.

The structure of the motor cycle—the sequence of visited energy minima during different stages of the rocking—is the unique signature of each device. The most interesting motor cycles generated at $z = 0$ in each of our three devices are presented in Fig. 4.39. Again, the average motor trajectory is plotted by red during the positive and in blue line during the negative phase of the rocking. The light gray trajectories show single stochastic realizations. The corresponding tension curves are shown in Fig. 4.40.

As we have already seen, in the hard device, the X-tilted is trapped in a single energy well of the double well potential and the power stroke element does not contribute to force generation. We can force the X-tilted ratchet to visit both minima of the double well potential if we use somewhat less realistic parameters, $k_0 = 7$, $k_1 = 7$, $l = 0.22$, $L = 2$, $\lambda = 1.4$, $\Phi_{max} = 5$ and $\alpha = 5$. Such cycle is formed only when the motor makes sufficiently large steps along the coordinate x and can therefore recharge the power stroke element. After a transient stage, an X-tilted ratchet with these parameters performs the cycle in the clockwise direction:

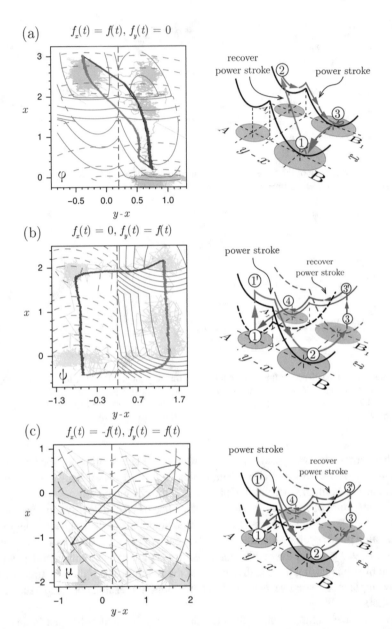

(a) $f_x(t) = f(t),\ f_y(t) = 0$

(b) $f_x(t) = 0,\ f_y(t) = f(t)$

(c) $f_x(t) = -f(t),\ f_y(t) = f(t)$

Fig. 4.39 The most complex motor cycles in hard device at $z = 0$: (**a**) X ratchet, (**b**) Y ratchet, (**c**) XY ratchet. The parameters $k_0 = 7,\ k_1 = 7,\ l = 0.22,\ L = 2,\ \lambda = 1.4,\ \Phi_{max} = 5,\ \tau = 20,$ $\alpha = 5$

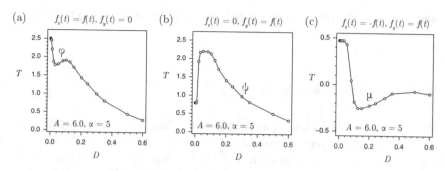

Fig. 4.40 Average tension T versus temperature D in: X ratchet (**a**), Y rathet (**b**) and XY ratchet (**c**). The parameters are the same as in figure above

$1 \rightarrow 2$. Due to the broken space symmetry, the motor advances in the x direction and crosses the energy barrier along the periodic potential $\Phi(x)$. In the meanwhile the motor recharges the power stroke element (performs the transition from the lower energy well B to the higher energy well A_1 in the bistable potential). During the next step $2 \rightarrow 3$, as the rocking force changes its sign, the new configuration of the energy landscape forces the motor into backward direction along the x axis. However, the motor is now trapped and instead of going backwards, it performs the power-stroke. Then the motor cycle starts again.

Consider now the Y-tilted ratchet. If we choose parameters as in the case of X-ratchet above we obtain the four-state cycle shown in Fig. 4.39b. The system reattached and jumps to a new cite on the actin filament while stretching the bi-stable spring. This behavior can be interpreted as a power-stroke (red path) followed by the recharging (blue path) in the attached state. The motor first goes through mechanical configurations A, B and then through the configurations B_1, A_1. In the ensuing four-state cycle the stages are: the transition $1' \rightarrow 2$ is the power stroke, then, as the motor makes a jump into the next nearest well along the actin filament (in the positive x direction) the rocking force changes its sign forcing the recharging of the power stroke mechanism, $3' \rightarrow 4$.

Note that in the corresponding two-state cycle obtained in the section on Y-tilted ratchets, the system was residing in a distant, force generating well of the periodic potential while performing periodic oscillations between the two conformational states of the power stroke element. The level of the generate force was high because the cross bridge was firmly attached throughout the cycle. In the four-state regime studied here, the system is periodically reaching the distant well of the periodic potential as well, but remains there only for a limited time before returning back to the original attachment site. In such regimes the average force is necessarily smaller.

Our Fig. 4.39c shows the typical cycle exhibited by the XY ratchet in the hard device. The motor periodically changes the attachment site: the cycle is performed between the state A_1 and the state A, B, corresponding to different attachment sites along the actin filament. In this regime the motor is able to generate the highest level of average tension. The three- state cycle can be interpreted as follows: first motor detachment and advance accompanied by the recharging of the power stroke device and then the power stroke combined with re-attachment brings the system into the original site.

4.6.3 Stochastic Energetics

We now apply to all devices the same conservative load $f_{ext} = -0.1$. In Fig. 4.41a we illustrate the temperature dependence of the mechanical efficiency for the X-tilted ratchet. In the regime of small amplitudes A we observe a maximum at finite temperature. The negative values of efficiency indicate the regimes where the motor is unable to perform a positive mechanic work against the external force and works instead as an active breaking mechanism. The Stokes efficiency, also shown in Fig. 4.41a, is always positive by definition.

We illustrate the efficiency of the Y-ratchet in Fig. 4.41b. Overall this ratchet is less efficient when the X ratchet. We can explain this difference by the design of the active mechanism: the metabolic energy is taken by the bi-stable element and therefore only partially consumed by the forward steps long the x direction.

Finally, the performance of the XY tilted ratchet is illustrated in Fig. 4.41c. At small amplitudes A we again observe a maximum of the mechanical efficiency at finite temperature. Interestingly, we find the XY device is the least efficient among all. One problem with our XY-tilted ratchet model is that it still interprets ambiguously the detached state which is present only implicitly. To deal with this conceptual problem we consider in the next section a more sophisticated model of the XY-tilted ratchet where we also take into consideration the explicit feedback between the state of the power stroke element and the degree of interaction between the myosin heads and the actin filament.

4.7 XY-Tilted Ratchet with a Steric Feedback

In this section we argue that the conformational state of the power-stroke element can provide steric regulation of the distance between the myosin head and the actin filament. More specifically, we assume that when the lever arm swings, the interaction of the head with the binding site weakens, see Fig. 4.42a. This and other aspects of steric rotation-translation coupling in ratchet models have been previously discussed in [43, 68, 90].

A schematic representation of the proposed model is shown in Fig. 4.42b, where x is the observable position of a myosin head, y is the internal variable characterizing the phase configuration of the power stroke element and z is another internal variable responsible for the coupling. The "macroscopic" variable x sees a symmetric energy landscape and is not directly affected by the ATP hydrolysis. Both asymmetry and driving can then originate only from the coupling between the external and the internal degrees of freedom.

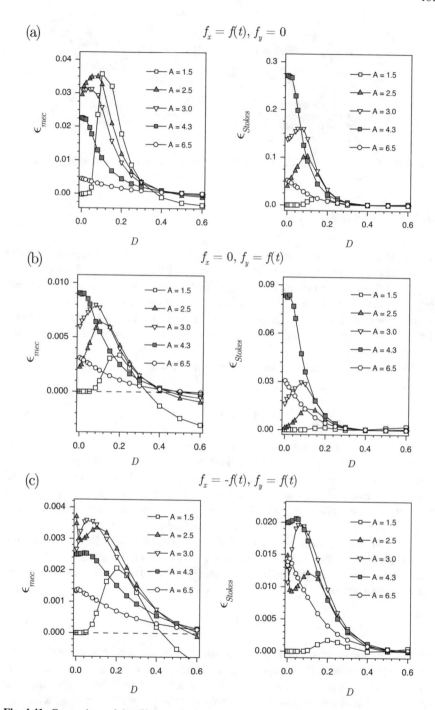

Fig. 4.41 Comparison of the efficiency in the three models loaded in the soft device with $f_{ext} = -0.1$; (**a**) X ratchet. (**b**) Y ratchet. (**c**) XY ratchet. The parameters $k_0 = 1.5$, $k_1 = 0.43$, $l = 0.3$, $L = 1$, $\lambda_1 = 0.7$, $V_{max} = 1.5$, $\alpha = 1$, $\tau = 30$

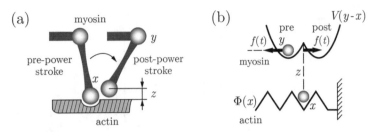

Fig. 4.42 (a) An illustration of the steric effect associated with the power-stroke; (b) sketch of the mechanical model

4.7.1 The Model

We identify the external degree of freedom with the variable x representing the location of actin binding face on the actin filament. We recall that the natural internal degree of freedom, describing the configurational state of the power-stroke element, is $y - x$, where the variable y was defined in the Introduction. By adding the second internal variable z, characterizing the separation of the myosin head and the actin filament, we attempt to capture the higher-dimensional effects of detachment in the simplest 1D setting.

The role of different variables is clear from the way we write the energy of the system

$$\hat{G}(x, y, z) = z\Phi(x) + V(y - x), \tag{4.26}$$

where $\Phi(x)$ is a non-polar periodic potential representing the binding strength of the actin filament and $V(y - x)$ is a double-well potential describing the power-stroke element. The two-well structure of the potential implies that the power-stroke mechanism can be either folded into the post-power-stroke state or unfolded into the pre-power-stroke state. For simplicity, we assume that the two wells of the potential $V(y - x)$ are symmetric which eliminates a redundant polarity.

The coupling between the state of the power-stroke element $y - x$ and the spatial position of the motor x is implemented through the internal variable z. In the simplest version of the model z is assumed to be a function of the state of the power-stroke element

$$z(x, y) = \Psi(y - x). \tag{4.27}$$

This function must have a particular structure in order to mimic the underlying steric interaction, see Fig. 4.43. We assume that when a myosin head executes the power-stroke it moves away from the actin filament and therefore the control function $\Psi(y - x)$ should progressively switch *off* the actin potential. Similarly, when the power-stroke is recharging the myosin head moves closer to the actin filament and the function $\Psi(y - x)$ should bring the actin potential back into *on* configuration.

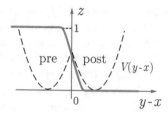

Fig. 4.43 The coupling function $\Psi(y-x)$ linking the degree of attachment z with the state of the power-stroke element $y - x$

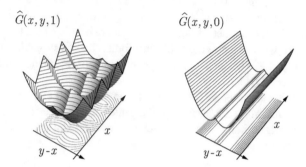

Fig. 4.44 The energy landscapes: $\hat{G}(x, y, 1)$, describing the attached state where $\Psi(y-x) = 1$, and $\hat{G}(x, y, 0)$, describing the detached state, where $\Psi(y-x) = 0$

We observe that since the double-well potential $V(y - x)$ is fully symmetric, the assignment of the wells to pre- or post-power-stroke states is arbitrary. Had we decided to invert the choice presented in Fig. 4.43 by relabeling the energy wells, we would have to replace $\Psi(s)$ by $\Psi(-s)$. As we see later in the paper, such switch results in a simple reversal of the directionality of the motion.

By using the coupling (4.27) we can simply eliminate the variable z and introduce the redressed potential

$$G(x, y) = \hat{G}(x, y, \Psi(y - x)). \qquad (4.28)$$

As it tracks the state of the power-stroke element the potential $G(x, y)$ effectively "flashes" between the periodic and flat (in x) configurations, see Fig. 4.44. However, in contrast to conventional flashing ratchets, the switch here is not imposed from outside but results from the coupling with a fluctuating internal variable.

The overdamped stochastic dynamics of the system with energy (4.28) is described by the following 2D system of (dimensionless) Langevin equations

$$\begin{cases} dx/dt = -\partial_x G(x, y) - f(t) + \sqrt{2D}\xi_x(t), \\ dy/dt = -\partial_y G(x, y) + f(t) + \sqrt{2D}\xi_y(t). \end{cases} \qquad (4.29)$$

Here $\xi(t)$ is a conventional white noise with $\langle \xi_i(t) \rangle = 0$, and $\langle \xi_i(t)\xi_j(s) \rangle = \delta_{ij}\delta(t-s)$. The parameter $D = k_B\theta/E$ is a dimensionless measure of temperature θ and k_B is the Boltzmann constant; for simplicity the viscosity coefficients are assumed to be the same for variables x and y. The force couple $f(t)$ with zero average represents a correlated component of the noise and characterizes mechanistically the degree of non-equilibrium in the external reservoir (the abundance of ATP).

We can say that the system (4.29) describes the power-stroke-driven ratchet because the correlated noise $f(t)$ acts on the relative displacement $y - x$. It effectively "rocks" the bi-stable potential and the control function $\Psi(y-x)$ converts such "rocking" into the "flashing" of the periodic potential $\Phi(x)$. Various other types of rocked-pulsated ratchet models have been previously studied in [99, 102].

The goal of any ratchet design is to generate a systematic drift $v = \lim_{t\to\infty}\langle x(t)\rangle/t$ without applying a biasing force. This is possible in the model governed by Eq. (4.29) because of an implicit symmetry breaking imposed by the control function (4.27).

To justify this claim, let us for simplicity set $f(t) = 0$ and rewrite (4.29) in the variables representing the position of the center of mass $q = (x + y)/2$ and the power-stroke configuration $r = y-x$, which is a conventional step in such problems [32]. The new potential is

$$G(q, r) = \Psi(r)\Phi(q - r/2) + V(r),$$

and if we recall that the equilibration of the variable r takes place at much faster time scale than the overall drift, we can adiabatically eliminate it and obtain a one dimensional stochastic system with an effective periodic potential

$$G_{eff}(q) \sim \ln\left[\int_{-\infty}^{\infty} \exp\left(-G(q, r)/D\right)dr\right].$$

In the absence of the feedback $\Psi(s) = 0$ this potential is symmetric $G_{eff}(q) = G_{eff}(-q)$ because $\Phi(s) = \Phi(-s)$ and $V(s) = V(-s)$. When $\Psi(s) \neq \Psi(-s)$, it loses symmetry because pre- and post-power-stroke configurations are no longer equivalent. It is also clear that by reverting the control function $\Psi(s) \to \Psi(-s)$, we change the directionality of the average motion, see Fig. 4.48 below.

To understand the dependence of the average velocity on the parameters of the model, we studied the system (4.29) numerically. In our computational illustrations we use again a periodic extension of the symmetric triangular potential $\Phi(x)$ with amplitude Φ_{max} and period L, see Fig. 4.42a

$$\Phi(x) = \begin{cases} \dfrac{2\Phi_{max}}{L}x & \text{if } 0 \leq x < L/2, \\[3mm] \dfrac{2\Phi_{max}}{L}(L - x) & \text{if } L/2 \leq x < L. \end{cases}$$

The symmetric potential $V(y - x)$ is assumed to be bi-quadratic with the same stiffness k in both phases. The distance between the bottoms of the wells is denoted by a, see Fig. 4.42b, so

$$V(y - x) = \begin{cases} \dfrac{1}{2}k(y - x + a/2)^2 & \text{if } y - x < 0, \\ \dfrac{1}{2}k(y - x - a/2)^2 & \text{if } y - x \geq 0. \end{cases}$$

The correlated component of the noise $f(t)$ is again interpreted as the simplest ac driving described by a periodic extension of a rectangular shaped function with amplitude A and period τ

$$f(t) = A(-1)^{n(t)} \quad \text{with} \quad n(t) = \lfloor 2t/\tau \rfloor,$$

where brackets $\lfloor \ \rfloor$ denote integer part. Finally, the steric control ensuring the gradual switch of the actin potential is described by a gradual step function

$$\Psi(s) = (1/2)\left[1 - \tanh(s/\varepsilon)\right], \tag{4.30}$$

where ε is a small parameter.

To fix the parametrization, we need to specify the dimensional scales. It is natural to use the distance between the bottoms of the wells in the bi-stable potential as the length scale l so $a = 1$. We have also made a standard assumption that the separation between the binding cites along the actin filament is of the same order as the power-stroke size and therefore $L = 1$. The height of the barrier between the binding sites was chosen as the energy scale E, so we put $\Phi_{max} = 1$. The relaxation time scale was set by the viscosity coefficient η and therefore $\tau^* = \eta l^2/E$. To ensure that the ac driving is slow at the scale of internal relaxation we took $\tau = 10$. The curvature of the energy wells in the bistable potential should be comparable with E/l^2 and therefore we took a generic value $k = 1.5$. In the computations we used the value of the small parameter $\varepsilon = 0.2$ which made the attachment and the detachment events sufficiently sharp.

To integrate the system (4.29) numerically, we used the simplest Euler–Maryama scheme [63] with a constant time step $\Delta t = 0.5 \times 10^{-3}$. The ensemble averaging was performed over $N = 10^4$ stochastic realizations.

Our numerical results are summarized in Fig. 4.45. First of all, we see that the drift is absent ($v = 0$) when the noise is uncorrelated and the external reservoir is in equilibrium ($A = 0$). This is an obvious consequence of the potential nature of this holonomic model. Indeed, the stationary probability flux satisfies $\nabla J = 0$ and $J = f\mathbf{F} - D\nabla f$ where $f(x, y)$ is the stationary probability distribution and $\mathbf{F}$ is the internal force. Since $\mathbf{F} = -\nabla G$, one can use periodicity in x and growth in $y - x$ (of the potential G) to show that $\mathbf{J} = 0$, see also [73, 93, 130].

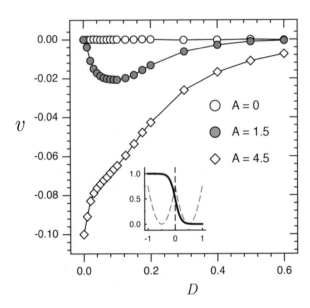

Fig. 4.45 The dependence of the average velocity v on temperature D and the amplitude of the ac signal A in the model with coupling (4.30). The pre- and post-power-stroke states are labeled in such a way that the purely mechanical ratchet would move to the left

It is then clear that the drift in this model is exclusively due to $A \neq 0$. When A is small, the drift velocity shows a maximum at finite temperatures which implies that the system exhibits stochastic resonance [41]. At high amplitudes of the ac driving, the motor works as a purely mechanical ratchet and the increase of temperature always worsens the performance [49, 58, 95].

As we have already seen, the direction of motion in this model is decided by the choice of steric biasing of the otherwise symmetric bi-stable potential. the chosen directionality can be either enhanced or suppressed if we consider polar actin filaments. To illustrate this point, we show in Fig. 4.46 how the drift velocity depends on the parameter characterizing the spatial asymmetry of the actin track. In particular, we see that on a polar filament with sufficient asymmetry our motor can be stopped and even steered in the opposite direction.

The next question concerns the compatibility of the proposed model with the minimal bio-chemical ATPase cycle shown in Fig. 4.1. The traditional identification of chemical and structural states, detailed in this figure, suggests that the motor must pass through the following four mechanical transients: "attached pre-power-stroke", "attached post-power-stroke", "detached post-power-stroke" and "detached pre-power-stroke". It is immediately clear that not all of these states can be reached by the model with coupling (4.30). Indeed, the detachment takes place when the "striking" element is positioned exactly between the two energy wells and therefore the power-stroke cannot be completed in the attached state. As a result, the model reproduces reliably only two structural configurations: the attached pre-power-stroke state and the detached post power-stroke state.

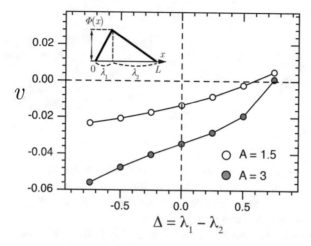

Fig. 4.46 The dependence of the drift velocity v on the filament polarity $\Delta = \lambda_1 - \lambda_2$ in the model with coupling (4.30) at fixed temperature $D = 0.01$

To capture the remaining states shown in Fig. 4.1 we must assume that the detachment, necessarily implying in our model the motion of the center of mass, is delayed till the power-stroke is (almost) completed. Similarly, the attachment must take place only after the power-stroke element has been (almost fully) recharged. The necessary modification of the model, accounting for such two-way delays, is discussed in Sect. 4.7.2.

4.7.2 Hysteretic Coupling

To reproduce the whole Lymn–Taylor cycle, we postulate that the switching of the actin potential from *on* to *off* state takes place at different values of the variable $y-x$, depending on the *direction* of the conformational change (folding or unfolding). To this end, we replace the holonomic coupling (4.27) by a memory operator

$$z\{x, y\} = \widehat{\Psi}\{y(t) - x(t)\} \tag{4.31}$$

whose output depends on whether the system is on the "striking" or on the "recharging" branch of the trajectory, see Fig. 4.47. Such memory structure can be also described by a rate independent differential relation of the form

$$\dot{z} = Q(x, y, z)\dot{x} + R(x, y, z)\dot{y}, \tag{4.32}$$

where the implied non-integrability makes the model non-holonomic. Indeed, if we introduce a vector variable $\mathbf{u} = (x, y, z)$, and neglect the time dependent external

Fig. 4.47 The hysteresis operator $\widehat{\Psi}\{y(t) - x(t)\}$ linking the degree of attachment z with the previous hystory of the power-stroke configuration $y(t) - x(t)$

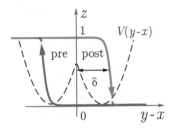

noise we can rewrite the system of the governing equations in the form $\dot{\mathbf{u}} = \mathbf{F}(\mathbf{u})$, where $\mathbf{F}$ is no longer a gradient. The resulting Brownian motor can potentially advance even in the absence of the correlated noise by extracting energy directly from the non-holonomic control mechanism.

By using (4.31) we can now rewrite the energy of the system as a functional of its history $y(t)$ and $x(t)$

$$G\{x, y\} = \widehat{\Psi}\{y(t) - x(t)\}\Phi(x) + V(y - x). \tag{4.33}$$

In the Langevin setting (4.29), the history dependence may mean that the underlying microscopic stochastic process is non-Markovian (due to, say, configurational pinning [14]), or that there are additional non-thermalized degrees of freedom that are not represented explicitly [50]. In general, it is well known that the realistic feedback implementations always involve delays [36].

To simulate hysteretic response numerically we used two versions of the same coupling function (4.30) shifted by δ with the branches $\Psi(y - x \pm \delta)$ identified sufficiently far away from the hysteresis domain, see Fig. 4.47. Our numerical experiments show that the performance of the model is not sensitive to the shape of the hysteresis loop and depends mostly on its width characterized by the small parameter δ.

In Fig. 4.48 we illustrate the "gait" of the motor with the hysteretic coupling (4.31). The center of mass advances in steps and during each step the power-stroke mechanism gets released and then gets recharged again, which takes place concurrently with attachment-detachment. By coupling the attached state with either pre- or post-power-stroke state, we can vary the directionality of the motion. The average velocity increases with the width of the hysteresis loop which shows that the motor can extract more energy from the coupling mechanism system with longer delays.

The results of the parametric study of the model are summarized in Figs. 4.49 and 4.50. First observe that the motor can move even in the absence of the correlated noise, at $A = 0$, because the non-holonomic coupling (4.33) breaks the detailed balance by itself. At finite A the system can use both sources of energy (hysteretic loop and ac noise) and the resulting behavior is much richer than in the non-hysteretic model.

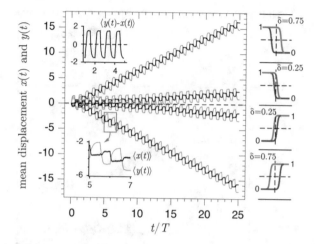

Fig. 4.48 Stationary particle trajectories in the model with the hysteretic coupling (4.31). Different ways of biasing lead to different directions of drift and large hysteresis loops produce faster moving motors. Other parameters are: $D = 0.02$ and $A = 1.5$

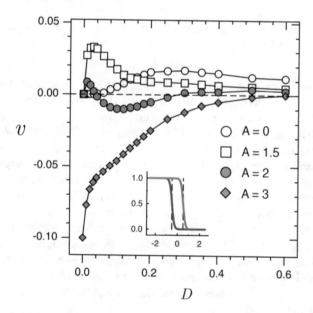

Fig. 4.49 The dependence of the average velocity v on temperature D in the hysteretic model with $\delta = 0.5$

For instance, if the holonomic ratchet with a fixed coupling bias always advances in one direction, the non-holonomic ratchet can self-propel in both directions. At large A the hysteretic motor exhibits the same directionality as the non-hysteretic motor and the average velocity is only mildly affected by the presence of the

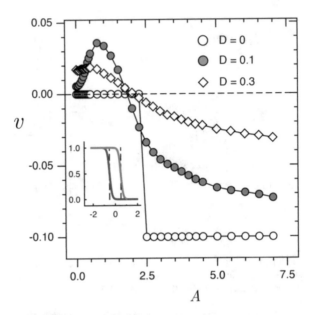

Fig. 4.50 The dependence of the average velocity v on amplitude of the ac driving A in the hysteretic model with $\delta = 0.5$

hysteresis. At small A the situation changes and now the direction of the drift is controlled by the hysteresis and is reversed comparing to the case of a non-hysteretic motor. As we see, in the hysteretic power-stroke-driven ratchet different active mechanisms dominate at different values of A. This opens an interesting possibility for these molecular machines to flip "engines" and in this way reverse the directionality by simply changing the intensity of the external energy supply.

The A dependence of the drift velocity is shown in more detail in Fig. 4.50. At zero temperature the system is pinned and the drift is blocked till the driving amplitude reaches a threshold beyond which the system can work as a mechanical ratchet. At finite temperatures the pinning disappears because of the noise-induced barrier crossing. At small A the motor drifts in the direction opposite to the direction of the mechanical ratchet. The velocity of this drift shows a characteristic peak at finite A revealing stochastic resonance. The current reversal, indicating the change of the mechanism from hysteresis-dominated to correlation-dominated, takes place near the depinning point $A \sim 2.5$.

To illustrate the mechanism of the hysteresis-dominated drift it is sufficient to consider the case when $A = 0$. This disables an alternative ac driven ratchet mechanism. In Fig. 4.51 we compare two realizations of particle trajectories in the 3D space $(x, y - x, z)$: for the model without hysteresis (4.27) and with hysteresis (4.31) obtained by projecting these trajectories on the 2D plane $(y - x, z)$ describe the structure of the corresponding "strokes" in the configurational space. In the holonomic case (4.27) the area of the projected loop is equal to zero and

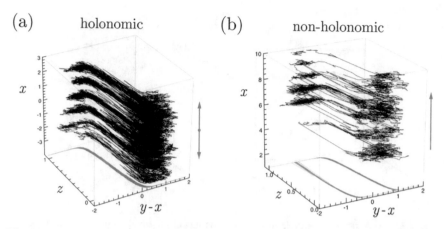

Fig. 4.51 Single particle trajectories in the space $\{y - x, z, x\}$ and their projections on the plane $(y-x, z)$: (**a**) non-hysteretic coupling (**b**) hysteretic coupling. The parameters are: $D = 0.2$, $A = 0$ and $\delta = 0.5$

we observe diffusion without drift (in Fig. 4.51a the average of x is equal to zero). Instead, in the non-holonomic case (4.31) the projected trajectory spans a finite area and the drift velocity is finite (see Fig. 4.51b). Similar dependence of the drift velocity on the area of the "stroke" is known in the theory of Stokes swimmers where non-holonomic control is also the factor responsible for the directional motion (in "violation" of the scallop theorem [4]).

The mechanical "stroke" in the space of internal variables $z, y - x$ can be compared with the minimal biochemical acto-myosin cycle shown in Fig. 4.1. The chemical states constituting this cycle are identified with structural configurations (obtained from crystallographic reconstructions) in the following way [72]: A(attached, pre-power-stroke $\rightarrow$ AM^*ADP^*Pi), B(attached, post-power-stroke $\rightarrow$ AM^*ADP), C(detached, post-power-stroke $\rightarrow$ M^*ATP), D(detached, pre-power-stroke $\rightarrow$ M^*ADP^*Pi). In our model the jump events are replaced by continuous transitions and the association of chemical states with particular regimes of stochastic dynamics is not straightforward.

In Fig. 4.52a, we show a fragment of the averaged trajectory of a steadily advancing motor projected on the $(x, y - x)$ plane. In Fig. 4.52b the same trajectory is shown in the $(x, y - x, z)$ space with fast advances in the z direction intentionally schematized as jumps. By using the same letters A, B, C, D as in Fig. 4.1 we establish a basic connection between the chemical/structural states and the transient mechanical configurations of the advancing motor.

Suppose that we start at point A corresponding to the end of the negative cycle of the ac driving $f(t)$. The system is in the attached, pre-power-stroke state and $z = 1$. As the sign of the force $f(t)$ changes, the motor undergoes a power-stroke and reaches point B while remaining in the attached state. When the configurational variable $y - x$ passes the detachment threshold, the myosin head detaches which

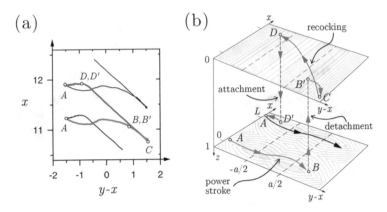

Fig. 4.52 (a) A steady-state cycle in the hysteretic model projected on the $x, y - x$ plane; red path if $f(t) > 0$ and blue path if $f(t) < 0$; (b) representation of the same cycle in the $z, x, y - x$ space with identification of the four chemical states A, B, C, D constituting the Lymn–Taylor cycle shown in Fig. 4.1. The level sets represent the energy landscape G at $z = 0$ (detached state) and $z = 1$ (attached state). The parameters are: $D = 0.02$, $A = 1.5$, and $\delta = 0.75$

leads to a transition from point B to B' on the plane $z = 0$. Since the positive cycle of the force $f(t)$ continues, the motor completes the power-stroke by moving from B' to point C. At this moment, the rocking force changes sign again which leads to recharging of the power-stroke mechanism in the detached state, described in Fig. 4.1 as a transition from C to D. In point D, the variable $y - x$ reaches the attachment threshold. The myosin head reattaches and the system moves to point D' where $z = 1$ again. The recharging continues in the attached state as the motor evolves from D' to a new state A, shifted by one period.

In this way the chemical states constituting the minimal enzyme cycle can be linked to the mechanical configurations traversed by our stochastic dynamical system. The detailed mechanical picture, however, looks more complicated than in the simplest Lymn–Taylor scheme. It is clear that at least in some regimes one can use the Kramers approximation to perform a controlled transition from our continuous dynamics to a description in terms of a discrete set of chemical reactions. However, it is also clear that more chemical states than in the minimal Lymn–Taylor model will be needed to describe the detailed mechanical "stroke".

So far we have been dealing with motors overcoming viscous friction but not carrying cargoes. The next step is to see how fast the same motor can move against an external force f_{ext}. Two different mechanical configurations of the motor carrying cargo correspond to the cases when $f_{ext} > 0, v < 0$ and $f_{ext} < 0, v > 0$, see Fig. 4.53. Since the non-hysteretic motor is designed to move to the left, the mechanical configuration shown in Fig. 4.53a can be somewhat arbitrarily characterized as "pushing". Given that the motor with the hysteretic coupling can move in both directions, the configuration shown in Fig. 4.53b also corresponds to a steady regime which can be then interpreted as "pulling". Since our motor does not have explicit leading and trailing edges, we assume that the force f_{ext} acts in both

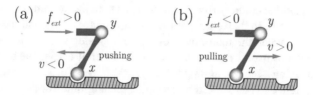

Fig. 4.53 Schematic representation of the power stroke driven motor carrying cargo: (**a**) pushing regime, (**b**) pulling regime

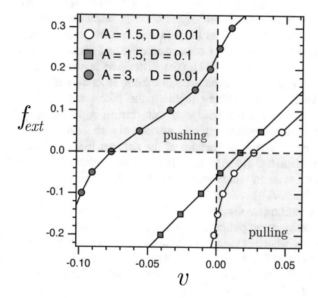

Fig. 4.54 The force-velocity relation in the model with hysteretic coupling at different amplitudes of the ac driving A and different temperatures D. The hysteresis width is $\delta = 0.5$

cases on the variable y which amounts to tilting of the potential (4.33) along the y direction

$$G\{x, y\} = \widehat{\Psi}\{y(t) - x(t)\}\Phi(x) + V(y - x) - yf_{ext}. \tag{4.34}$$

However, the actual architecture of a half sarcomere is asymmetric and the forces are transmitted through passive cross-linkers imposing a particular polarity on the loading. Therefore, despite the ambiguity, we find the association of the two mechanical regimes shown in Fig. 4.53 with pushing and pulling appropriate.

A stochastic system with energy (4.34) was studied numerically and in Fig. 4.54 we show the computed force-velocity relations. The light quadrants in the (f_{ext}, v) plane correspond to two domains of dissipative behavior where $R = f_{ext}v > 0$. Here the direction of the force agrees with the direction of motion and the motor is being dragged by the applied load (while exhibiting both passive and active friction).

The shaded quadrants indicate the two domains where the system is anti-dissipative and $R = f_{ext}v < 0$. In these regimes the motor produces work and the motion can be of two types: when the motor overcomes the opposing pushing force and drives the cargo ahead of itself ($f_{ext} > 0$, $v < 0$) and when it carries the cargo attached from behind acting against a pulling force ($f_{ext} < 0$, $v > 0$). Since in the hysteretic model the current can be reversed by changing the amplitude of the ac noise A, our motor can perform two types of useful work.

Observe that at low temperatures the convexity properties of the force-velocity relations in active pushing and active pulling regimes are different. In the case of pulling the typical force-velocity relation is reminiscent of the Hill's curve describing isotonic contractions [51]. In the case of pushing, the force-velocity relation can be characterized as convex-concave and such behavior has been also observed in muscles [30, 31, 71]. The difference is due to the dominance of physically non-equivalent mechanisms in the corresponding parameter domains.

For instance, in the pushing regimes, the motor activity fully depends on ac driving and at large amplitudes of this driving it performs as a mechanical ratchet. Instead, in the pulling regimes, associated with small amplitudes of external driving, the motor advances because of the delayed feedback exemplified by the hysteretic mechanism. We may speculate that both mechanisms can be operative in acto-myosin systems which would then provide an explanation for occasionally counterintuitive drift directions.

We also mention that dissimilarity of convexity properties of the force-velocity relations in pushing and pulling regimes has been recently discussed in the context of cell motility where acto-myosin contractility is known to be one of the main driving forces [94]. The direct quantitative comparison is, however, premature since in our minimal setting the model deals with a single cross bridge and still neglects important collective effects [19].

4.7.3 Non-potential Models

The performance of the power-stroke driven ratchet can be considerably enhanced if the feedback between the power-stroke and the attachment-detachment mechanisms is made non-conservative even in the absence of hysteresis. This would happen, for instance, if the configurational state of the power-stroke element affected the position of a myosin head with respect to actin filament, while the reverse influence remained insignificant, in other words, if the coupling between the power-stroke element and the actin potential was one-sided. In this case instead of a passive control we are dealing with an *active control* represented by a Maxwell demon-type mechanism [18, 35].

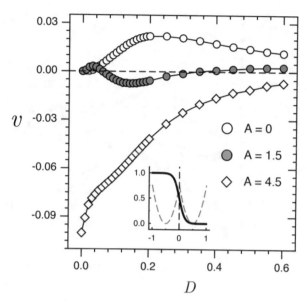

Fig. 4.55 Temperature dependence of the drift velocity v in the non-potential model (4.35) without hysteresis

The governing equations describing such ratchet can be written in the form

$$\begin{cases} dx/dt = -z\partial_x\Phi(x) - \partial_x V(y-x) - f(t) + \sqrt{2D}\xi_x(t) \\ dy/dt = -\partial_y V(y-x) + f(t) + \sqrt{2D}\xi_y(t), \end{cases} \quad (4.35)$$

where the notations are the same as in (4.29). The results of the numerical study of the system (4.35) are summarized in Fig. 4.55.

The overall behavior of the non-potential system (4.35) is similar to the behavior of the potential system with hysteretic coupling (4.33). Since the ratchet can now receive energy from the active controlling device [73, 93], a nonzero drift takes place already at $A = 0$. The direction of the current can be again reversed by varying the amplitude of the driving. At large values of A, we obtain our usual mechanical ratchet which does not see the non-potentiality of the model. At small A the ratchet exploits the non-potentiality of the model in the essential way. As in the case of hysteretic system, the direction of the drift is now opposite to the one picked up by the mechanical ratchet. Notice also that at moderate values of A the directionality of the drift can be reversed by the variation of temperature.

The non-potential ratchet shows the highest performance in combination with the hysteretic feedback (4.33), see Fig. 4.56. The behavior of such hybrid system at $A = 0$ is similar to what we have seen in the case of the system with energy (4.33) which means that in this regime the response is dominated by hysteresis. As A increases we observe a new effect: around $A \sim 1.5$ the system appear to be in

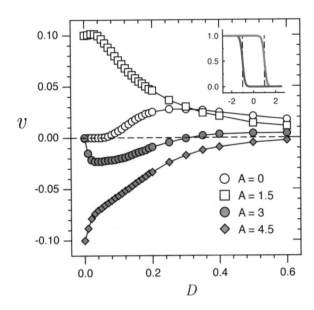

Fig. 4.56 Temperature dependence of the drift velocity v in the non-potential model (4.35) with hysteresis, $\delta = 1$

a resonant state and works as a quasi-mechanical ratchet, however now, the non-potentiality is the principle driving factor, see Fig. 4.56. With further increase of A we observe a reversal of the current and the system enters the regime where the main driving force is again the ac noise. At large A the mechanical ratchet behavior prevails again, however, it is fundamentally different from the quasi-mechanical ratchet behavior observed around $A \sim 1.5$.

In Fig. 4.57 we illustrate the effect of the amplitude A on the drift velocity in more detail. In contrast to the potential case, the ratchet can now move at zero temperatures in both directions equally fast if the amplitude of the ac signal is chosen appropriately. The current reversal takes place in the narrow range of amplitudes A where the transition from a mechanical to a quasi-mechanical ratchet mechanism takes place.

At finite temperatures we see a complex interplay of all three active mechanisms. The detailed study of the underlying stochastic system, allowing one to precisely map the parametric domains where particular mechanisms dominate, will be presented elsewhere.

To better understand the effects of non-potentiality we also compute the Péclet number $Pe = Lv/D_e$, characterizing the relative strength of the drift (over diffusion). The effective diffusion coefficient is defined by Reimann et al. [97], Lindner et al. [70], and Khoury et al. [61]

$$D_e = \frac{1}{2} \lim_{t \to \infty} \frac{\langle [x(t) - \langle x(t) \rangle]^2 \rangle}{t}, \tag{4.36}$$

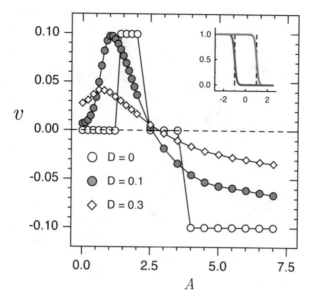

Fig. 4.57 The dependence of the average velocity v on the amplitude A of the ac signal in the non-potential model with hysteresis, $\delta = 1$

so the stochastic transport is most coherent when the absolute value of the Péclet number is larger than one. From Fig. 4.58 we see that only in the non-potential model the motion at small values of the driving amplitude A can be viewed as truly directional.

Suppose now that a load is attached to the motor with non-potential hysteretic coupling. The typical force-velocity relations are shown in Fig. 4.59. As in the potential case, the motor can operate in two anti-dissipative regimes either by working against a pushing force or by pulling a cargo. At both small and large values of A the behavior of the potential and the non potential motors is similar. Expectedly, an anomaly takes place in the pulling regime ($f_{ext} < 0$, $v > 0$) at $A \sim 1.5$ where the motor behaves as a quasi-mechanical ratchet. Here the non-potentiality dominates and the force-velocity relation shows an unusually sharp convexity change. It is interesting that in this regime the behavior near the stall force is reminiscent of the one observed in skeletal muscles [53].

4.8 Active Rigidity

In this section we show that effective rigidity or, more generally, effective susceptibility in a bundle of elastically coupled cross-bridges, can emerge from the activity localized at the level of the power stroke machinery. Consider again a skeletal muscle cell [53] where we neglect detachment of active cross-linkers (cross-

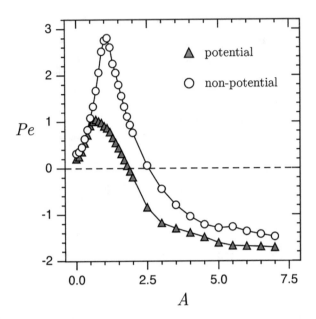

Fig. 4.58 The Péclet number in the potential model with hysteresis ($\delta = 0.5$ as in Fig. 4.50) and in the non-potential model with hysteresis ($\delta = 1$ as in Fig. 4.57); $D = 0.1$

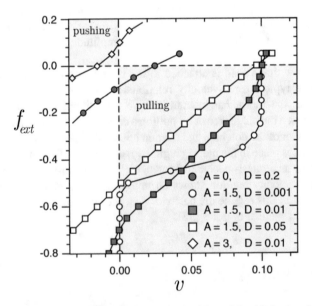

Fig. 4.59 The force-velocity relation in the non-potential model with hysteresis at different temperatures D and different driving amplitudes A; $\delta = 1$

bridges) and model an elementary series element (half-sarcomere) as a parallel array of molecular motors operating in stall conditions. As in other sections, we model attached myosin motors as bi-stable springs, with two energy wells corresponding to pre and post power stroke configurations. Each 'snap-spring' of this kind acts against a linear spring, representing a structural filament. The system is exposed to both uncorrelated agitation (scaled with temperature-type parameter D) and a correlated noise representing ATP hydrolysis (scaled with affinity-type parameter A).

4.8.1 Macroscopic Problem

We start with an assumption that a muscle myofbril can be viewed as a chain of half-sarcomeres arranged in series with each half-sarcomere represented by a parallel array of N cross-bridges interacting with a single actin filament [54, 120], see Fig. 4.60. We assume again that the nontrivial dynamics of attached cross-bridges is due exclusively to the conformational change in myosin heads (power stroke) and model cross-bridges as bi-stable elements in series with linear springs, see Fig. 4.61. We therefore stay with our original paradigm that the nonequilibrium driving is provided through the rocking of the bi-stable elements [108].

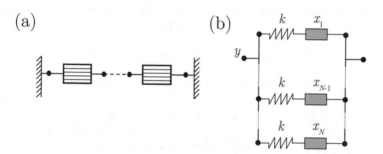

Fig. 4.60 (a) Schematic representation of a muscle myofibril as a series connection of half sarcomeres; (b) Model of a single half-sarcomere with attached cross-bridges arranged in parallel. Shaded boxes in (b) represent bi-stable snap-springs shown in Fig. 4.61

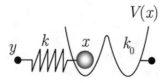

Fig. 4.61 Schematic representation of a bi-stable snap-spring in series with a linear spring

A half-sarcomere in this model, see Fig. 4.60b, can be described by the system of non-dimensional Langevin equations

$$\begin{cases} dx_i/dt = -\nabla_{x_i}\Phi + \sqrt{2D}\xi(t), \\ \nu dy/dt = -\nabla_y\Phi, \end{cases} \qquad (4.37)$$

where y is a macro-scopic variable characterizing the strain at the level of the half-sarcomere whose dynamics is slow due to the large value of the relative viscosity ν. The variable y is coupled with N fast soft-spin type variables x_i through identical springs with stiffness k. The potential energy is $\Phi = \sum_{i=1}^{N} E(x_i, y, t) - f_{ext}y$, where $f_{ext}(t)$ is a slowly varying macro-scopic force. The ensuing problem is a soft spin generalization [76] of the Huxley-Simmons model [54] and we recall that its applications extend far beyond muscles mechanics, from hair cell gating [77] and binding of cell-adhesion patches [33] to mechanical denaturation of RNA and DNA hairpins [128] and unzipping of biological macromolecules [47].

The equation for y in (4.37) can be re-written as

$$\frac{\nu}{N}\frac{dy}{dt} = k\left(\frac{1}{N}\sum_{i=1}^{N}x_i - y\right) + \frac{f_{ext}}{N}, \qquad (4.38)$$

which makes the mean field nature of the interaction between y and x_i explicit. If N is large, we can replace $\frac{1}{N}\sum_{i=1}^{N}x_i$ by $\langle x \rangle$ using the fact that the variables x_i are identically distributed and exchangeable [34]. If $\nu_0 = \nu/N$ and $g_{ext} = f_{ext}/(kN)$ remain finite in the limit $N \to \infty$, we can write

$$\nu_0 dy/dt = k[(\langle x \rangle - y) + g_{ext}(t)].$$

Assume for determinacy that the function $f(t)$ is periodic and choose its period τ in such a way that $\Gamma = \nu_0/k \gg \tau$. Since $g_{ext}(t)$ is a slowly varying function at the time scale τ, we can split the force $k(\langle x \rangle - y)$ acting on y into a slow component $k\psi(y) = k(\overline{\langle x \rangle} - y)$ which originates from our effective potential and a slow-fast component $k\phi(y,t) = k(\langle x \rangle - \overline{\langle x \rangle})$ which in the steady regime becomes a τ periodic function of time with zero average. We can then write

$$\Gamma\frac{dy}{dt} = \psi(y) + \phi(y,t) + g_{ext}. \qquad (4.39)$$

The next step is to average (4.39) over the time scale τ. To this end we introduce a decomposition $y(t) = z(t) + \zeta(t)$, where z is the averaged (slow) part of the motion and ζ is a fast varying perturbation (with time scale τ) that is small compared to z. Then, expanding 4.39 up to first order in ζ, we obtain,

$$\Gamma\left(\frac{dz}{dt} + \frac{d\zeta}{dt}\right) = \psi(z) + \partial_z\psi(z)\zeta + \qquad (4.40)$$

$$\phi(z,t) + \partial_z\phi(z,t)\zeta + g_{ext}.$$

Since $g_{\text{ext}}(t) \simeq \tau^{-1} \int_t^{t+\tau} g_{\text{ext}}(u)du$, we obtain at fast time scale [13]

$$\Gamma \frac{d\zeta}{dt} = \phi(z, t).$$

Integrating this equation between t_0 and $t \le t_0 + \tau$ we can assume that z is fixed and therefore $\zeta(t) - \zeta(t_0) = \Gamma^{-1} \int_{t_0}^t \phi(z(t_0), u)du$. Given that ϕ is τ periodic with zero average, we conclude that $\zeta(t)$ is also τ periodic with zero average.

If we now average (4.40) over the fast time scale τ, we obtain

$$\Gamma dz/dt = \psi(z) + r + g_{\text{ext}},$$

where

$$r = (\Gamma\tau)^{-1} \int_0^\tau \int_0^t \partial_z\phi(z, t)\phi(z, u)dudt.$$

Since both $\phi(z, t)$ and $\partial_z\phi(z, t)$ are bounded, we can write $|r| \le (\tau/\Gamma)c \ll 1$, where the 'constant' c depends on z but not on τ and Γ. Therefore, if $N \gg 1$ and $v/(kN) \gg \tau$, the equation for

$$z(t) = \tau^{-1} \int_t^{t+\tau} y(u)du$$

can be written directly in terms of the effective potential

$$(v/N)dz/dt = -\partial_z F + f_{\text{ext}}/N.$$

To find the potential $F(z)$ we need to average the ensuing mean field model over the fast and slow-fast dynamics in (4.41) while keeping the variable y fixed.

4.8.2 Mean Field Model

The implied mean field model can be viewed as a description of a probe characterized by a (microscopic) coordinate x which is placed in an active environment. The probe is attached through an elastic spring to a measuring device associated with a (meso-scopic) variable y. We assume that the variable y is slow and treat it as a control parameter. Instead, the variable $x(t)$ will undergo fast stochastic motion which will have to be averaged out.

In the absence of noise, the environment will be introduced through the potential $V(x)$ and we assume that the probe is placed in an unstable configuration. One way to satisfy this condition is to assume that $V(x)$ has a double well structure with the reference position of a probe in a spinodal state. We further assume that the probe is exposed to a fluctuating surrounding medium with a quickly relaxing component represented by an equilibrium thermostat and a relatively slower relaxing component describing non-equilibrium environment. We study the meso-scopic force exerted by the probe on the measuring device which implies the transition from the environment potential $V(x)$ to the effective potential for the measuring device $F(y)$.

To be more specific, consider the stochastic dynamics of a variable $x(t)$ described by a dimensionless Langevin equation

$$dx/dt = -\partial_x E(x, y, t) + \sqrt{2D}\xi(t), \tag{4.41}$$

where $\xi(t)$ is a standard white noise and D is a temperature-like characteristic of the equilibrium thermostat. The potential $E(x, y, t) = E_p(x, t) + E_m(x, y)$ is sum of two components: $E_p(x, t) = V(x) - xf(t)$, describing the probe in an out of equilibrium environment and $E_m(x, y) = k(x - y)^2/2$, describing the linear elastic coupling with a measuring device characterized by stiffness k. We assume that the energy is supplied to the system through the rocking force $f(t)$ with zero average which is characterized by an amplitude A and a time scale τ. To have analytical results, we need to further assume that the potential $V(x)$ is bi-quadratic

$$V(x) = (|x| - 1/2)^2/2. \tag{4.42}$$

Similar framework has been used before in the studies of directional motion of molecular motors [25].

To compute the effective potential $F(y)$ we use an observation that if the 'measurements' are performed at a time scale larger that τ, the resulting force is $T(y) = k[y - \langle x \rangle]$, where the averaging is over ensemble and time

$$\langle x \rangle = \lim_{t \to \infty} (1/t) \int_0^t \int_{-\infty}^{\infty} xp(x, t)dxdt.$$

Here $p(x, t)$ is the probability distribution for the variable x which solves the associated Fokker–Plank equation. The primitive of the averaged tension

$$F(y) = \int^y T(s)ds, \tag{4.43}$$

can be then viewed as a non-equilibrium analog of the free energy [6, 66, 100, 129]. While in our case, the mean-field nature of the model ensures potentiality of the averaged tension, in a more general setting, the averaged stochastic forces will lose their gradient structure and even the effective "equations of states" relating averaged forces with the corresponding generalized coordinates may not be well defined [9, 12, 44, 56, 110, 114].

It is clear that the effective potential $F(y)$ will depend not only on $V(x)$ but also on the stochastic properties of the driving $f(t)$. The question we pose is whether there exists a non-biased stochastic driving which ensures stabilization of spinodal configurations that would be unstable in the absence of the noise. In the equilibrium case, when $f(t) = 0$, such stabilization is possible because of entropic effects but only at sufficiently large temperature D. The challenge is to find a correlated (colored) noise $f(t)$ which ensures stabilization at arbitrary small D. The possibility of bi-modality of the marginal probability distribution $p(x, t)$ in single-well potentials is known for DC and Levi type noises [28, 29], however, this effects disappears after ensemble averaging involved in the computation of the effective potential $F(y)$.

4.8.3 Non-dimensionalization

Equation (4.41), which constitutes the basis of our prototypical model, is dimensionless. To translate the results back into muscles context we need to use the time scale $\tau^* = \eta/k_0 \sim 0.1$ ms where $\eta \sim 0.38$ ms. pN/nm is the micro-scale viscosity [19] and $k_0 \sim 3$ pN/nm is the passive stiffness of the equivalent energy wells. The spatial scale is then $l^* = a$ where $a \sim 10$ nm is the distance between two minima of the pre and post power stroke wells [69] and the energy scale is $\epsilon^* = k_0 a^2 \sim 300$ pN $\cdot$ nm.

Following [19] we also assume that $k = k_m/k_0 \sim 0.6$, where $k_m \sim 2$ pN/nm is the stiffness of the elastic part of the myosin motor [7, 67]. Hence $D = k_B \Theta/(k_0 a^2) \sim 0.01$ where $k_B = 4.10$ pN $\cdot$ nm is the Boltzmann constant, $\Theta \sim 300$ K is the ambient temperature, and $a = 10nm$ is the characteristic size of a motor power-stroke [69]. For the active driving we obtain $\tau = \tau_a/(\eta/k_0) \sim 100$ where $\tau_a = 40$ ms is the characteristic time of ATP hydrolysis [53]. We can now write that $A = \sqrt{\Delta\mu/(k_0 a^2)} \approx 0.5$ where $\Delta\mu = 20k_B\Theta$ is the typical value of degree of non-equilibrium in terms of the affinity of ATP hydrolysis reaction [53].

The knowledge of the set of dimensionless parameters A, D and τ will be sufficient to locate the muscle system on the phase (regime) diagram. Such diagrams will be constructed in Sect. 4.8.4 for three different types of active driving.

4.8.4 Phase Diagrams

In this Section we consider the general problem (4.41) at finite temperature ($D > 0$) when both equilibrium and nonequilibrium reservoirs are contributing to the microscopic dynamics simultaneously. The limiting case of zero temperatures ($D = 0$) will be analyzed separately in Sec. 4.8.5.

Periodic (P) Driving Suppose first that the non-equilibrium driving is represented by a periodic (P), square shaped external force $f(t) = A(-1)^{n(t)}$ with $n(t) = \lfloor 2t/\tau \rfloor$, where brackets denote the integer part. While this choice of periodic driving ensures certain analytical simplicity, the obtained results will be generic.

It will be convenient to rewrite the dynamic equation (4.41) in the form

$$dx/dt = -\partial_x \tilde{V}(x, z) + f(t) + \sqrt{2D}\xi(t), \qquad (4.44)$$

where

$$\tilde{V}(x, z) = \frac{1}{2}(|x| - 1/2)^2 + \frac{1}{2}k(x - z)^2.$$

The associated Fokker–Planck equation for the time dependent probability distribution $p(x, t)$ reads

$$\partial_t p = \partial_x [p\partial_x E(x, t) + D\partial_x p]. \qquad (4.45)$$

First of all we note that explicit solution of (4.45) can be found in the adiabatic case when the correlation time τ is much larger than the escape time for the bistable potential V [48, 74]. The idea of this approximation is that the time average of the steady state probability can be computed from the mean of the stationary probabilities with constant driving force (either $f(t) = A$ or $f(t) = -A$).

It is obvious, that the adiabatic approximation becomes exact in the special case of an equilibrium system with $A = 0$ when the stationary probability distribution is known explicitly:

$$p_0(x) = Z^{-1}e^{-\tilde{V}(x)/D},$$

where $Z = \int_{-\infty}^{\infty} \exp(-\tilde{V}(x)/D)dx$. The tension elongation curve can then be computed analytically, since we know

$$\overline{\langle x \rangle} = \langle x \rangle = \int_{-\infty}^{\infty} xp_0(x)dx. \qquad (4.46)$$

The resulting curve $T(z)$ and the corresponding potential $F(z)$ are shown in Fig. 4.62a. At zero temperature the equilibrium system with $A = 0$ exhibits negative stiffness at $z = 0$ where the effective potential $F(z)$ has a maximum (spinodal state). As temperature increases we observe a standard entropic stabilization of the configuration $z = 0$, see Fig. 4.62a.

Computing solution of the equation $\partial_z T|_{z=0} = 0$, we find an explicit expression for the critical temperature $D_e = r/[8(1 + k)]$ where r is a root of a transcendental equation $1 + \sqrt{r/\pi}e^{-1/r}/[1 + \mathrm{erf}(1/\sqrt{r})] = r/(2k)$. The behavior of the roots of the equation $T(z) = -k(\langle x \rangle - z) = 0$ at $A = 0$ is shown in Fig. 4.63b. It illustrates a second order phase transitions taking place at $D = D_e$.

In the case of constant force $f \equiv A$ the stationary probability distribution is also known [98]

$$p_A(x) = Z^{-1}e^{-(\tilde{V}(x) - Ax)/D},$$

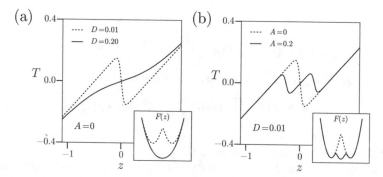

Fig. 4.62 Tension elongation curves $T(z)$ in the case of periodic driving (adiabatic limit). The equilibrium system ($A = 0$) is shown in (**a**) and out-of-equilibrium system ($A \neq 0$)—in (**b**). The insets show the effective potential $F(z)$. Here $k = 0.6$

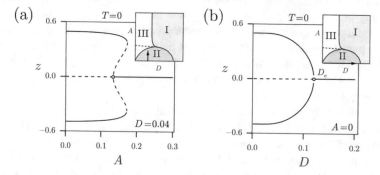

Fig. 4.63 The parameter dependence of the roots of the equation $T(z) = 0$ in the adiabatic limit: (**a**) fixed $D = 0.04$ and varying A, first order phase transition (line $C_A - M_A$ in Fig. 4.64a); (**b**) fixed $A = 0$ and varying D, second order phase transition (line $D_e - C_A$ in Fig. 4.64a). The dashed lines correspond to unstable branches. Here $k = 0.6$

where again $Z = \int_{-\infty}^{\infty} \exp(-\tilde{V}(x)/D)dx$. In adiabatic approximation we can write the time averaged stationary distribution in the form, $p_{Ad}(x) = \frac{1}{2}[p_A(x) + p_{-A}(x)]$, which gives

$$\overline{\langle x \rangle} = \frac{1}{2}[\langle x \rangle(A) + \langle x \rangle(-A)]. \tag{4.47}$$

In this equation the expression for $\langle x \rangle(A)$ can be written explicitly

$$\langle x \rangle(A) = Z^{-1} \sum_{i=1,2} P(u_i)[\sqrt{\pi} u_i \operatorname{erfc}(u_i) - (-1)^i e^{-u_i^2}],$$

where

$$P(u) = (D/(1+k))e^{-\frac{1}{2D}\left(\frac{1}{4}+kz^2-2Du^2\right)},$$

$$u_{1,2} = (A \pm 1/2 + kz)/\sqrt{2D(1+k)},$$

$$Z = \sqrt{(1+k)\pi/(2D)} \sum_{i=1,2} P(u_i)\mathrm{erfc}((-1)^i u_i),$$

and erfc is the complementary error function.

The force-elongation curves $T(z)$ and the corresponding potentials $F(z)$ obtained for $A \neq 0$ are shown in Fig. 4.62b. It demonstrates the main effect: as the degree of non-equilibrium, characterized by A, increases, not only the stiffness in the state $z = 0$ where the original double well potential V had a maximum changes from negative to positive but also the effective potential $F(z)$ develops around this point the third well. We interpret this phenomenon as the emergence of active rigidity because the new equilibrium state becomes possible only at a finite value of the driving parameter A while the temperature parameter D can be arbitrarily small. The behavior of the roots of the equation $T(z) = -k(\langle x \rangle - z) = 0$ at $A \neq 0$ is shown in Fig. 4.63a. It illustrates the first order phase transitions taking place at increasing A (and small fixed D).

The full steady state regime map (dynamic phase diagram) summarizing the results obtained in adiabatic approximation is presented in Fig. 4.64a. There, the 'paramagnetic' phase I describes the regimes where the effective potential $F(z)$ is convex, the 'ferromagnetic' phase II is a bi-stability domain where the potential $F(z)$ has a double well structure and, finally, the 'Kapitza' phase III is where the function $F(z)$ has three convex sections separated by two concave (spinodal) regions. Note that the boundaries of the domain occupied by phase III in this diagram are not defined by the number of the roots of $T(z) = 0$, as it is usually done

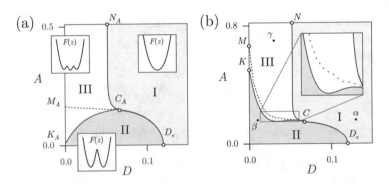

Fig. 4.64 Phase diagram in (A, D) plane showing phases I,II and III: (**a**) adiabatic limit, (**b**) numerical solution at $\tau = 100$ (**b**). C_A is the tri-critical point, D_e is the point of a second order phase transition in the passive system. The "Maxwell line" for a first order phase transition in the active system is shown by dots. Here $k = 0.6$

in the study of magnetic systems, but by the counting of the number of the effective "energy wells" linked to convexity properties of the whole effective potential $F(z)$.

In view of the structure of the bifurcation diagrams shown in Fig. 4.63, we can interpret the boundary $C_A - D_e$ separating phases I and II as a line of (zero force) second order phase transitions and the dashed line $C_A - M_A$ as a Maxwell line for the (zero force) first order phase transition, see Fig. 4.63. Then C_A can be interpreted as a tri-critical point near which the system can be described by an non-equilibrium (active) Landau potential of the form

$$F(z) = F_0 + rz^2 + qz^4 + pz^6,$$

where r, q, p are pseudo-thermodynamic parameters. Indeed, while r represents the usual measure of temperature D and $p > 0$ is a constant, the A dependent parameter q is an uncoventional measure of the intensity of active driving. Similar tri-critical point appears in the periodically driven mean field Suzuki-Kubo model of magnetism which can be interpreted in our terms as a description of the zero tension behavior [115].

The adiabatic approximation fails at low temperatures (small D) where the escape time diverges and in this range the phase diagram has to be corrected numerically, see Fig. 4.64b. By simulating directly Eq. (4.41) we obtain that even the moderate temperature features of the diagram (tri-critical point, point D_e and the vertical asymptote of the boundary separating phases I and III at large values of A) are captured adequately by the adiabatic approximation. For instance, the value of temperature corresponding to point N (at infinite A) obtained from the adiabatic approximation is $D_N = q/[8(1 + k)]$ where q is a solution of a transcendental equation $q - k = q^{3/2}/[\sqrt{q} + e^{1/q}\sqrt{\pi}(1 + \text{erf}(1/q))]$ which agrees with our numerics.

The new feature of the non-adiabatic phase diagram is a dip of the boundary separating Phases II and III at some $D < D_e$ leading to an interesting re-entrant behavior (cf. [89, 118]). This is an effect of stochastic resonance which is not captured by adiabatic approximation.

To verify our numerical results in the low temperature domain $D \to 0$ we used Kramers approximation, valid when the rocking period τ is much smaller than the typical escape time of the bi-stable potential V.

It allows one to compute explicitly the location of point K ($A = 1/2$) and point M ($A = 1/2 + k/4$), which we found to be in full agreement with our numerical simulations, see Fig. 4.64b. Because of incompatibility of the limits $D \to 0$ and $\tau \to \infty$ these points are rather far from the corresponding adiabatic predictions K_A and M_A shown in Fig. 4.64a.

Force-elongation relations characterizing the mechanical response of the system at different points on the (A, D) plane (Fig. 4.64b) are shown in Fig. 4.65 where the upper insets illustrate the typical stochastic trajectories and the associated cycles in $\{\langle x(t)\rangle, f(t)\}$ coordinates. We observe that while in phase I thermal fluctuations dominate periodic driving and undermine the two wells structure of the potential, in phase III the jumps between the two energy wells are fully synchronized with the

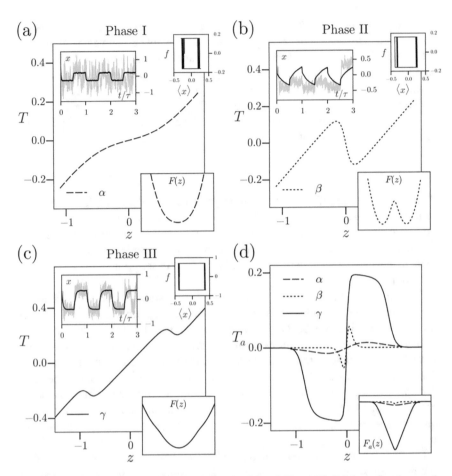

Fig. 4.65 (**a–c**) Typical tension-length relations in phases I, II and III. Points α, β and γ are the same as in Fig. 4.64b; (**d**) shows the active component of the force. Inserts show the behavior of stochastic trajectories in each of the phases at $z \simeq 0$ (gray lines) superimposed on their ensemble averages (black lines); the stationary hysteretic cycles, the structure of the effective potentials $F(z)$ and the active potential $F_a(z)$ defined as a primitive of the active force $T_a(z)$. The parameters: $k = 0.6$, $\tau = 100$

rocking force. In phase II the system shows intermediate behavior with uncorrelated jumps between the wells.

In Fig. 4.65d we illustrate the active component of the force $T_a(z) = T(z; A) - T(z; 0)$ in phases I, II and III. A salient feature of Fig. 4.65d is that active force generation is significant only in the resonant (Kapitza) phase III. A biologically beneficial plateau (tetanus) is a manifestation of the triangular nature of a pseudo-well in the active landscape $F_a(z) = \int^z T_a(s)ds$; note also that only slightly bigger $(f, < x >)$ hysteresis cycle in phase III, reflecting a moderate increase of the extracted work, results in considerably larger active force. It is also of interest that the largest active rigidity is generated in the state $z = 0$ where the active force is equal to zero.

If we now estimate the non-dimensional parameters of the model by using the data on skeletal muscles, see Sect. 4.8.3, we obtain $A = 0.5, D = 0.01, \tau = 100$. This means that muscle myosins in stall conditions (isometric contractions), may be functioning in resonant phase III. Our simple model can therefore contribute to the explanation of the observed stability of skeletal muscles in the negative stiffness regime [19]; similar mechanism may be also behind the titin-based force generation at long sarcomere lengths [103].

Dichotomous (DC) Driving The P driving is only one among correlated signals that can serve as a mechanical representation of an out of equilibrium chemical reservoir. To ascertain the robustness of the results obtained in the case of P driving we now consider another type of correlated forcing which is also characterized by two parameters, the amplitude A and the characteristic time τ. It is given by the explicit formula $f(t) = A(-1)^{n(t)}$, where $n(t)$ is a Poisson process with $P(n) = e^{-\lambda}\lambda^n/n!$ with $\lambda = 1/(2\tau)$ and is known as symmetric dichotomous (DC) noise or random telegraph signal [55, 86]. For this Markov process we have $\langle f(t) \rangle = A \exp(-t/\tau)$ and $\langle f(t), f(s) \rangle = A^2 \exp(-|t-s|/\tau)$.

The probability distribution can be written in the form $p(x, t) = p_-(x, t) + p_+(x, t)$ where $p_{\pm}(x, t)$ are the probability densities to be in a state x at time t given that $f = \pm A$. The DC driven system (4.41) is described by the two coupled Fokker–Planck equations [10],

$$\partial_t p_{\pm} = \partial_x(\partial_x E_{\pm} p_{\pm} + D\partial_x p_{\pm}) + \lambda(\mp p_{\pm} \pm p_{\mp}) \tag{4.48}$$

where $E_{\pm}(x) = \tilde{V}(x) \mp Ax$. Note that in this interpretation the DC noise appears as a chemical reaction violating the detailed balance [91]. The stationary version of the system (4.48) can be written in a transparent form if in addition to $p(x) = p_-(x) + p_+(x)$ we introduce a complimentary variable $d(x) = p_+(x) - p_-(x)$. Then we obtain

$$\begin{aligned} \partial_x \tilde{V} p - D\partial_x p - Ad &= 0, \\ \tau \partial_x(\partial_x \tilde{V} d - D\partial_x d - Ap) &= d. \end{aligned} \tag{4.49}$$

The numerical study of (4.41) with DC noise shows that the qualitative structure of the phase diagram in the (A, D) plane remains the same as in the case of P driving, see Fig. 4.66. We again observe phases I, II and III and the tri-critical point at about the same location as in the case of P noise.

To interpret the numerical results, it is instructive to consider analytically tractable special cases. First of all, Eq. (4.49) can be used to obtain the adiabatic $(\tau \to \infty)$ limit when the two equations decouple and the steady state probability distributions take the form $p_{\pm}(x) \sim \exp(-E_{\pm}(x)/D)$ as in the case of P driving. The resulting phase diagrams are therefore identical, see section "Periodic (P) Driving".

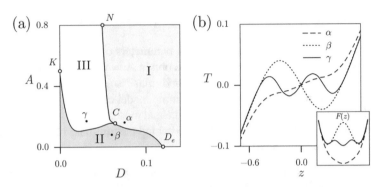

Fig. 4.66 (a) Phase diagram in the case of DC driving. The identification of phases I, II, and III is the same as in Fig. 4.64a, b. (b) Typical tension-length relations in different phases (b). Here $\tau = 100$ and $k = 0.6$

The second case, when the analytic results are available, is the zero temperature limit $D \to 0$ considered in detail in Sect. 4.8.5.

Finally, the third analytically tractable case is $\tau \to 0$, $A \to \infty$, with $\tilde{D} = A^2 \tau$ remaining finite. In this limit we obtain that the non-equilibrium component of the noise is represented by a Gaussian white noise $f(t) = \sqrt{2\tilde{D}}\xi_a(t)$ with the temperature $\tilde{D}$ that is different from the temperature of the equilibrium reservoir D, for instance, one can think about a system exposed to a thermostat with temperature D and a chemostat with temperature $\tilde{D}$. The combined excitations are again represented by a white noise $\sqrt{2D^*}\xi_n(t)$ with effective temperature $D^* = \sqrt{D^2 + \tilde{D}^2}$.

In contrast to the zero temperature case, now the Kapitza phase III, describing active stabilization, is absent. We obtain only phases I and II separated by a second order phase transition line $\sqrt{D^2 + \tilde{D}^2} = D_e$ with the universal asymptotic behavior $\tilde{D} \sim (D_e - D)^{1/2}$ near equilibrium, see Fig. 4.67. The system in this limit can

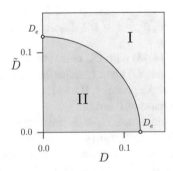

Fig. 4.67 Phase diagram for the case when the chemical reaction is modeled by an effective temperature $\tilde{D}$. Here $k = 0.6$

undergo entropic stabilization only which means that the two temperature model does not capture the same range of phenomena as the $D = 0$ model. Note that other two temperature models can exhibit destabilization of a single well system[44].

Ornstein–Uhlenbeck (OU) Driving We have seen that the overall effect of the two bounded noises on a mechanical system may be similar even though one of them is highly correlated and non-Markovian and another one is weakly correlated and Markovian. To show that not all noises are 'mechanically equivalent' we now consider an Ornstein–Uhlenbeck (OU) process which is also characterized by two parameters A and τ [8, 86].

In the case of OU driving, the function $f(t)$ is a solution of the stochastic equation

$$df(t)/dt = -\frac{1}{\tau} f(t) + A\sqrt{\frac{2}{\tau}} \xi_f(t). \tag{4.50}$$

Exactly as in the case of the DC noise we have for the first two moments $\bar{f}(t) = \langle f(t) \rangle = A \exp(-t/\tau)$, and $\langle f(s) f(t) \rangle = A^2 \exp(-|t - s|/\tau)$, where we assumed for determinacy that $f(0) = A$. The resulting process is also Markovian, however now it is unbounded and is defined on a continuous state space.

The Fokker–Planck equation for the probability density $p(x, f, t)$ takes the form

$$\partial_t p = \partial_x (p \partial_x E + D \partial_x p) + \tau^{-1} \partial_f (fp + A^2 \partial_f p). \tag{4.51}$$

Our numerical results for the system driven by OU noise are summarized in Fig. 4.68a. At small intensity of driving A we observe the conventional picture of entropic stabilization. A striking feature of this diagram is the absence of phase III, which means that in contrast to the cases of P and DC driving, the OU driven system does not support the phenomenon of active stabilization. To understand these

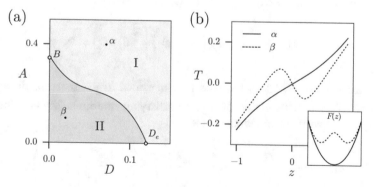

Fig. 4.68 (a) Phase diagram in the case of OU driving. The identification of phases I, II is the same as in Fig. 4.64a, b. (b) The typical tension-length relations in different phases. Here $\tau = 100$ and $k = 0.6$

numerical results it is instructive to consider the already mentioned three limiting cases that can be treated analytically.

In the adiabatic limit ($\tau \to \infty$), Eq. (4.51) simplifies and can be integrated. Then we obtain that $p(x) \sim \exp(-E(x)/D)$ which shows that in this limit only entropic stabilization remains possible.

Another analytically tractable limit is $D \to 0$, which shows again that in contrast to the cases of P and DC driving, only phases I and II are present at zero temperature phase diagram.

Finally, we can consider the double limit $\tau \to 0$, $A \to \infty$, with $\tilde{D} = A^2 \tau$ fixed. As in the case of DC noise, we recover in this limit a system subjected to an effective temperature and showing phases I and II only, see Fig. 4.67.

The analysis of these special cases supports our numerical results suggesting that in the OU driven system the tri-critical point is absent. We can link the failure to generate active rigidity in such system with the unbounded nature of the OU noise allowing the eventual escape from a neighborhood of any resonant state.

4.8.5   Zero Temperature Limit

To understand better the differences between our three representations of non-equilibrium driving, we now compare the behavior of the system in the analytically tractable limit when the temperature of an equilibrium thermostat is equal to zero, $D = 0$. In this limit the role of passive stabilization is minimized, which allows one to make the effect of active terms more transparent. When $D = 0$ we are left with two non-dimensional parameters: the correlation time τ and the amplitude of the active signal A. We found, however, that using another pair $(\tau, \tilde{D})$, with $\tilde{D} = A^2 \tau$, is more convenient.

Dichotomous (DC) Driving In the case of DC driving with $D = 0$ the stationary solution of the Fokker–Plank equation (4.49) can be written in the form [64]

$$p(x)\partial_x \tilde{V}(x) + A^2 \left[\frac{1}{\tau} - \partial_x \left(\partial_x \tilde{V}(x) \cdot \right) \right]^{-1} \partial_x p(x) = 0, \qquad (4.52)$$

where the notation $\partial_x \left(\partial_x \tilde{V} \cdot \right)$ should be understood in the sense of differential operators. The formal solution of ((4.52)) satisfying zero boundary conditions at infinity can be written in quadratures [48, 64]

$$p(x) = \frac{Z^{-1}}{A^2 - (\partial_x \tilde{V}(x))^2} \exp\left(-\frac{1}{\tau} \int^x \frac{\partial_y \tilde{V}(y)}{A^2 - (\partial_y \tilde{V}(y))^2} dy \right),$$

where we still need to find the normalizing constant Z. For this solution to be valid we must also satisfy the inequality

$$|\partial_x \tilde{V}(x)| < A. \tag{4.53}$$

When $A = 0$, we recover the deterministic case where condition (4.53) selects between the points $x_{0,1}(0)$ where the force vanishes.

In principle, the choice depends on the initial condition but in the limit of vanishing D and large time t, the trajectory $x(t)$ converges to the point minimizing the potential $\tilde{V}$. The resulting tension elongation relation can be then obtained by setting

$$\overline{\langle x \rangle} = \frac{x_0(0) + x_1(0)}{2} + \text{sign}(z)\frac{x_1(0) - x_0(0)}{2}.$$

The effective energy $F(z)$ emerges as a symmetric two parabolic bi-stable potential where $z = 0$ is a singular spinodal point separating the energy wells at $z = \pm 1/2$.

Another simple case is when $\tau \to 0$ with $A^2\tau = \tilde{D}$ remaining finite. In this limit activity disappears and driving becomes equilibrium with temperature $\tilde{D}$. The steady state probability distribution is given by $p(x) \sim \exp(-\tilde{V}(x)/\tilde{D})$ and the effective energy exhibits a transition from phase II to phase I at the critical temperature D_e.

To compute $p(x)$ in the general case, we identify the admissible set, where ((4.53)) holds, as $]x_0(-A), x_0(A)[\bigsqcup]x_1(-A), x_1(A)[$ where

$$x_0(-A) = \min(0, \tfrac{-1/2+kz-A}{1+k}) \le x_0(A)$$
$$x_0(A) = \min(0, \tfrac{-1/2+kz+A}{1+k}) \le 0$$
$$x_1(-A) = \max(0, \tfrac{1/2+kz-A}{1+k}) \ge 0$$
$$x_1(A) = \max(0, \tfrac{1/2+kz+A}{1+k}) \ge x_1(-A)$$

We can now integrate $p(x)$ on each of the segments $]x_0(-A), x_0(A)[$ and $]x_1(-A), x_1(A)[$. The result can be written in the form

$$p(x) = C_0\Psi_0(x)^{(2\tau(1+k))^{-1}-1}\mathbf{1}_{]x_0(-A),x_0(A)[}(x)$$
$$+ C_1\Psi_1(x)^{(2\tau(1+k))^{-1}-1}\mathbf{1}_{]x_1(-A),x_1(A)[}(x), \tag{4.54}$$

where $\Psi_0(x) = A^2 - [(1+k)x - kz + 1/2]^2$ and $\Psi_1(x) = A^2 - [(1+k)x - kz - 1/2]^2$.

If the domain of definition is connected as in, say, *Case 2*, when $x_0(A) = x_1(-A) = 0$, a continuity condition relates C_0 and C_1:

$$C_0 = Z^{-1}\Psi_1(0)^{(2\tau(1+k))^{-1}},$$
$$C_1 = Z^{-1}\Psi_0(0)^{(2\tau(1+k))^{-1}}. \tag{4.55}$$

If instead either $x_0(A)$ or $x_1(-A)$ is different from zero, the two sets $]x_0(-A), x_0(A)[$ and $]x_1(-A), x_1(A)[$ are separated by a segment where the probability is equal to zero. This means that the passage from one region to the other is impossible. In this case the coefficients C_0 and C_1 depend on the initial probability distribution as in the periodic case (at $D = 0$).

If we regularize the problem by adding a weak white noise (small $D \neq 0$), the choice of constants becomes again unambiguous as we can associate the support of the distribution with the side (0 or 1) opposite to the smallest potential barrier. We can then write explicitly $C_1 = Z^{-1}\max(0, \text{sign}(z))$ and $C_0 = Z^{-1} - C_1$. In all cases the constant Z is found from normalization.

We illustrate the stationary probability distributions $p(x)$ in Fig. 4.69a for several choices of parameters. The analytical expression for the tension elongation curves $T(z)$ involve hypergeometric functions and is too complex to be presented here. The resulting curves are illustrated in Fig. 4.69 for small and large values of the correlation time. The phase diagram, shown in Fig. 4.70a, exhibits all three phases I, II and III with a tri-critical point C' located at $\tau_{C'} = [2(k + 1)]^{-1}$ and $\tilde{D}_{C'} = D_e + [2(k + 1)]^{-1}/4$. The behavior of the force-elongation relations in different phases is illustrated in Fig. 4.70b. As we see, the DC driven dynamics is sufficiently rich to capture both active and entropic stabilization phenomena even in the absence of the equilibrium reservoir (at $D = 0$).

Periodic (P) Driving The numerical simulations for the problem with P driving and $D \to 0$ show only phases II and III even for rapidly oscillating external fields, see Fig. 4.70c.

To understand this result we can use Kramers approximation which can be developed under the assumption that the rocking period is short comparing to at least one of the escape times $\tau_{0,1}(\pm A)$. The use of such anti-adiabatic limit is consistent with the observation that in the limit $D \to 0$ the escape times from the energy wells diverge.

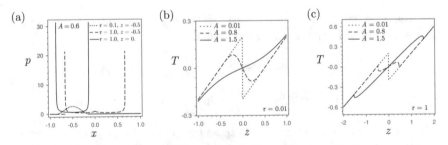

Fig. 4.69 (a) Examples of stationary probability distributions in the case of DC driving with $A = 0.6$. Dotted line: $\tau = 0.1$, $z = -0.5$. Dashed line: $\tau = 1$, $z = -0.5$. Solid line: $\tau = 1$, $z = 0$. (b–c) Tension elongation relations for different values of τ

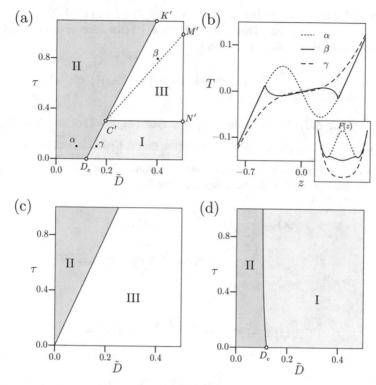

Fig. 4.70 (a) Zero temperature phase diagram in the case of DC driving. The identification of phases I, II and III is the same as in Fig. 4.64a, b. Tension-elongation relations in the case of DC driving in different phases (b). Zero temperature phase diagram in the case of P driving (c) and OU driving (d). Parameters $k = 0.6$, $D = 0$

A study of the purely mechanical problem with P driving reveals that, since the potential E can have up to four local minima, the dynamical system 4.44 can have up to four stationary solutions. We have four main cases:

- *Case 1:* $\overline{\langle x \rangle} = [x_0(-A) + x_0(A)]/2$
- *Case 2:* $\overline{\langle x \rangle} = [x_0(-A) + x_1(A)]/2$
- *Case 3:* $\overline{\langle x \rangle} = [x_1(-A) + x_1(A)]/2$
- *Case 4:*

$$\overline{\langle x \rangle} = \begin{cases} \frac{1}{2}[x_0(-A) + x_0(A)], & \text{if } z < 0 \\ \\ \frac{1}{2}[x_1(-A) + x_1(A)], & \text{if } z > 0. \end{cases}$$

To justify, for instance, the last expression (*Case 4*) we can apply the rate theory for rocked bi-stable system. Then we obtain for $n_0(t)$ (the number of particle in the well 0 at the moment t) the expression

$$\langle n_0 \rangle \sim e^{\left(-\min_{\pm A} [\Delta E_1(\pm A)] + \min_{\pm A, 1, 0} [\Delta E_{0,1}(\pm A)] \right)/D}.$$

Here either $\min_{\pm A} \Delta E_1(\pm A) = \min_{\pm A, 1, 0} \Delta E_{0,1}(\pm A)$ and $\langle n_0 \rangle = 1$, or $\min_{\pm A} \Delta E_1(\pm A) > \min_{\pm A, 1, 0} \Delta E_{0,1}(\pm A)$ and $\langle n_0 \rangle = 0$. The condition $\min_{\pm} \Delta E_1(\pm A) > \min_{\pm, 1, 0} \Delta E_{0,1}(\pm A)$ introduces the dependence of the stationary distribution on z. After time averaging, the steady state probability distribution takes the form:

$$p(x) = \frac{\langle n_0 \rangle}{2} [p_0(x; -A) + p_0(x; A)]$$

$$+ \frac{1 - \langle n_0 \rangle}{2} [p_1(x; -A) + p_1(x; A)], \tag{4.56}$$

where

$$p_0(x; \pm A) = \frac{\exp(-E_0(x; \pm A)/D)}{\int\limits_{-\infty}^{0} \exp(-E_0(x; \pm A)/D)\, dx}$$

$$p_1(x; \pm A) = \frac{\exp(-E_1(x; \pm A)/D)}{\int\limits_{0}^{\infty} \exp(-E_1(x; \pm A)/D)\, dx}.$$

In the limit $D \to 0$ the distributions $p_{0,1}(x; \pm A)$ become delta functions concentrated at the points $x_{0,1}(\pm A)$ which gives our formula for $\langle x \rangle$.

If we now use the computed values for $x_{0,1}(\pm A)$, we can obtain the analytic expressions for the tension $T(z)$. Then, by solving the equation $T(z) = 0$ we can locate the line of the first order phase transition separating phases II and III and show that $A = \frac{1}{2}$ at point K and that $A = \frac{1}{2}\left(1 + \frac{k}{2}\right)$ at point M, both in agreement with the numerical phase diagram presented in Fig. 4.71c. The qualitative difference between the predictions of the adiabatic approximation implying that D is large and the Kramers approximations corresponding to small D is illustrated in Fig. 4.71.

In coordinates $(\tau, \bar{D})$ the phase diagram for the P driven system with zero temperature shows a single phase boundary separating phases II and III, see Fig. 4.70c. The entropically stabilized phase II is absent because, despite the presence of the noise, it is bounded and there is no stochastic contribution allowing the system to cross arbitrary barriers. Because of the same reason the phase boundary between phases II and III in the P driven system is shifted comparing to the case of DC driving as the point D_e does not exist any more. This is in contrast to the fact that at finite D the two systems (with P and DC driving) behave quite similarly, in particular, they are indistinguishable in the adiabatic limit $\tau \to \infty$.

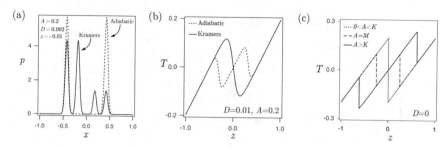

Fig. 4.71 (a) Examples of the stationary probability distributions in the case of P driving in adiabatic (dotted line) and Kramers (solid line) approximations at $A = 0.2$, $z = -0.01$, $D = 0.002$. (b) Tension elongation relations for $D = 0.01$ and $A = 0.2$ expressed by Kramers solution (solid line) and adiabatic solution (dotted line). (c) Tension elongation relations in the limit $D \to 0$ for several value of A. For $0 < A < K$, all curves collapse (dotted line) since the energy injected by the rocking is not sufficient to overcome the potential barriers

Ornstein–Uhlenbeck (OU) Driving In the case of OU driving with $D = 0$ an analytical approximation of the stationary probability distribution is available when $\tau << 1$ [48]. The main idea is to combine Eq. (4.44) and Eq. (4.50) to obtain a new equation for a noisy inertial oscillator

$$\frac{d^2x}{dt^2} + \frac{dx}{dt}(1 + \tau\partial_{xx}\tilde{V}(x)) + \partial_x\tilde{V}(x) = A\sqrt{2\tau}\xi_f(t), \qquad (4.57)$$

where ξ_f is a standard white noise. At large times the inertial dynamics with additive noise (4.57) can be approximated by the overdamped dynamics with multiplicative noise

$$\frac{dx}{dt} = (1 + \tau\partial_{xx}\tilde{V}(x))^{-1}\left(-\partial_x\tilde{V}(x) + A\sqrt{2\tau}\xi_f(t)\right)$$

which must be interpreted in the Stratanovitch sense [48]. The corresponding Fokker Planck equation

$$\partial_t p = \partial_x\left(\frac{\partial_x\tilde{V}}{1 + \tau\partial_{xx}\tilde{V}}p\right)$$
$$+ \partial_x\left(\frac{1}{1 + \tau\partial_{xx}\tilde{V}}\partial_x\left(\frac{A^2\tau}{1 + \tau\partial_{xx}\tilde{V}}p\right)\right), \qquad (4.58)$$

has an explicit stationary solution [48]:

$$p(x) = Z^{-1}|1 + \tau\partial_{xx}\tilde{V}(x)|\exp\left(-\frac{\tilde{V}(x) + \tau(\partial_x\tilde{V}(x))^2/2}{A^2\tau}\right).$$

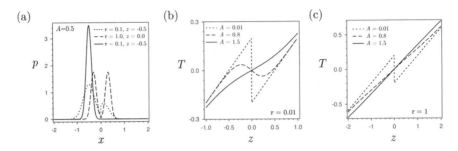

Fig. 4.72 (**a**) Examples of stationary probability distributions in the case of OU driving at $A = 0.5$; dotted line: $\tau = 1, z = -0.5$; dashed line: $\tau = 1, z = 0$; solid line: $\tau = 0.1, z = -0.5$. (**b–c**) Tension elongation relations in the OU cases for small (**b**) and large (**c**) correlation times. Large correlation times are formally outside of the domain of validity of the approximation

Notice again that when $\tau \to 0$ with $\tilde{D} = \tau A^2$ fixed, $f(t)$ becomes a white noise and the distribution $p(x)$ takes the classical Boltzmann form.

In Fig. 4.72a we show examples of the stationary distributions for specific values of parameters. The corresponding tension curves $T(z)$ are illustrated in Fig. 4.72b, c for large and small correlation times.

Our Fig. 4.70d shows the resulting phase diagram which, as expected, exhibits only phases I and II. This is again a confirmation of the fact that in the case of OU driving the crucial phase III, describing the phenomenon of active regidity, is absent. When τ is small (at a fixed $\tilde{D}$), the OU noise becomes a white noise and, as in the case of the DC driving, the phase boundary separating phases I and II passes through the point D_e.

The comparison of all three phase diagrams, shown in Fig. 4.70a, c, d suggests, that at zero temperature the system with DC driving is an intricate *amalgam* of the systems with OU and P drivings.

4.9 Conclusions

In these lecture notes we discussed the possibility that acto-myosin contraction is driven exclusively by the power-stroke. We developed several mechanistic models built on the assumption that the microscopic stochastic dynamics can be described by a set of continuous equations of mechanics with the equilibrium thermal environment modeled by a uncorrelated noise and internal activity modeled in different ways, in particular, by a correlated noise. The correlations associated with nonthermal such noise would then reflect the nonequilibrium nature of the external reservoir.

To model the full cross-bridge mediated actin-myosin interaction we developed three-dimensional phase space framework coupling a periodic potential with a bistable potential. In this perspective, the periodic potential represents myosin/actin interaction; the conformational change responsible for the power stroke is described by a double-well potential. The mechanical approach allowed us to reveal the important difference between the soft and hard device loading conditions. We mention that this asymmetry remained under-appreciated in the conventional chemo-mechanical models.

Our starting point was the observation that in the currently accepted mechanistic representations of acto-myosin systems, the power-stroke is undermined as a passive folding-unfolding mechanism while the attachment-detachment is given a primary role as the main driver of contraction. Since active sites are located inside motor domains, the external forces, representing the ATP activity, are typically introduced as conjugates to macroscopic positions of these domains and such ratchets are essentially attachment-detachment-driven. The implied mechanisms may be indeed operative during muscle contractions but then they would be complimentary to the ones studied here. In our approach the thrust of the ATP activity was shifted towards the internal variable characterizing the state of the power-stroke element.

Depending on the particular sub-unit where the external correlated force is applied, we introduced three basic designs of the force generating ratchet machinery. By localizing the effect of the correlated rocking on a single internal degree of freedom, we defined three basic models: X-tilted, Y-tilted and XY-tilted ratchets.

The X-tilted ratchet is the conventional mechanism where the external activity is concentrated in the actin filament. We have shown that with X-tilted ratchet design one cannot simulate the full four-state Lymn–Taylor cycle because the detachment of the cross-bridge head and the recharge of the power stroke are always combined. Another shortcoming of this model is that it does not treat adequately the detached configuration of the acto-myosin system.

In the Y-tilted ratchet model the correlated noise acts on the internal variable located inside the power stroke mechanism making both the power stroke and the actin filament active. The resulting ratchet performs four-state cycle in the soft device and either two-state or four-state cycle in the hard device. This suggests that the Y-tilted ratchet framework is capable in principle of mimicking the complete Lymn–Taylor cycle, however the mechanistic interpretation of such internal drive in term of the molecular structure of the cross-bridge remains ambiguous.

Finally, the XY-tilted ratchet model can be viewed as a mechanistic system which is driven entirely through the activity concentrated in the power stroke element while the actin filament is interpreted as passive. By treating the power stroke as the primary mechanism we delegated to the attachment/detachment process the secondary role of a machinery securing translational character of the motion. However, since the XY-tilted ratchet was shown to generate only three-state cycle it remains fundamentally incompatible with the existing biochemical models.

The main limitation of all these models is that the detached state is represented as a maximum of the periodic and therefore the detachment takes place very quickly. To overcome this problem we developed a synthetic model where the XY tilted ratchet mechanism was augmented by taking into consideration the explicit steric separation of thick and thin filaments. To make a clear distinction between our model and the conventional models of Brownian ratchets we assumed that the actin track is non-polar and that the bi-stable element is unbiased. The symmetry breaking was then achieved exclusively through the coupling of these two sub-systems.

Our synthetic model accounting for steric effects was based on the assumption that the multiplicative feedback is acting on the space-periodic potential $\Phi(x)$. In this model the conformational state of the power stroke mechanism regulates the distance of the myosin head form actin filament. We associated the pre-power stroke with a detached state (no interaction with the spatial periodic potential $\Phi(x)$) and the post power stroke—with the attached state (the system interacts strongly with the space periodic potential $\Phi(x)$). In this way the detached state was fully integrated into the mechanical cycle. The resulting model reproduced all four states of the Lymn–Taylor cycle where the individual states were interpreted as transient mechanical configurations.

By considering three classes of models of this type we exposed three different ways of how a power-stroke-driven ratchet can receive energy of the ATP hydrolysis and presented mechanical representations of the associated non equilibrium chemical reservoirs. In the first, traditional, representation, a mechanical action of the chemical reaction was modeled by a correlated component of the noise. The second representation was based on the idea that the coupling between internal and external degrees of freedom is hysteretic. Here in contrast to what is usually observed in macroscopic systems, hysteresis was used as a source rather than a sink of energy. The third representation implied that the internal degrees of freedom have an inherently chemical origin and therefore the source of non-equilibrium is in the lack of potentiality of these forces. We have shown that only the hysteretic design allows one to reproduce fully adequately the complete four state Lymn–Taylor cycle.

In the last section of these lecture notes we assumed that attachment detachment machinery is disabled and addressed the question whether a power stroke driven molecular device can generate effective rigidity. Instead of a single stall state, the proposed model was shown to exhibit a family of stall states and we quantified the energetic cost of moving from one member of the family to another. Since in our case the implied parameter had the meaning of meso-scopic strain, the derivative of the (time averaged) energy with respect to this parameter could be interpreted as the effective rigidity.

Our prototypical mean field model of active rigidity supports the idea that by controlling the degree of non-equilibrium in the system, one can stabilize apparently unstable or marginally stable mechanical configurations and in this way modify the structure of the effective energy landscape (when it can be defined). Our analysis, however, reveals that apparently similar noises can generate qualitatively different mechanical effects and that the very possibility of the power-stroke driven stabilization of an unstable state may be ultra-sensitive to the higher moments of the stochastic forces.

To summarize, we provided compelling evidence that a relatively simple mechanical system is able to generate complex stochastic dynamics imitating muscle contractions. In particular, we showed that such system can mechanistically reproduce the complete Lymn–Taylor cycle. This opens a way towards structural interpretation of the chemical states, known from the studies of the catalytic cycle in solution, and determining functionality of the particular transient mechanical configurations of the acto-myosin complex. The implied identification is a precondition for the bio-engineering reproduction of a wide range of cellular processes, where myosin cross-bridges play the dominant role, from movement of cells to cytokinesis. Given that the mechanisms involved in our model can be mimicked artificially at a super-cellular scale, the proposed schematization of the contraction phenomenon can be viewed as a step towards building engineering devices imitating acto-myosin enzymatic activity.

It is also important to mention that starting from the existing approach of rocking ratchets and reinventing it in the framework of the power stroke activity, we were able to unify the description of a single processive molecular motor such as Kinesin, with the description of the collectively operating non-processive molecular motors such as myosin. In this way we built a bridge between theoretical description myosin and Kinesin motors that have so far been treated as fundamentally different. In support of the idea that both processive and non-processive motors can be driven through a conformational change, we mention that the general shape of the force-velocity relations obtained in this paper is compatible with the available measurements not only for non-processive motors but also for processive motors [27, 60, 80, 123].

We showed, in particular, that while the most realistic XY tilted ratchet can perform a positive mechanical work, it is less efficient than X and Y ratchets. To understand why such seemingly inefficient device may be selected by evolution, it is important to remember that alternative, more efficient strategies include mechanical activity of actin filaments which is mechanically rather ambiguous.

The main limitation of the discussed picture of contraction is that it was developed for a single cross-bridge while important collective effects were neglected, see Fig. 4.73. A theory accounting for the implied collective effects has been so far developed only for passive response of skeletal muscles involving mechanically induced power stroke but not the attachment-detachment [19] and in active regimes one can expect a variety of interesting phenomena from coherent fluctuations [26, 45, 59, 121] to self-tuning towards criticality [5, 106]. Yet another reason for

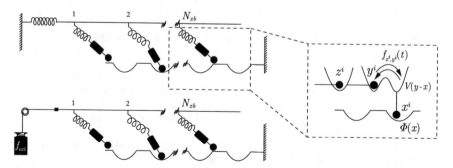

Fig. 4.73 Schematic representation of collectively interacting myosin motors: (**a**) hard device configuration. (**b**) soft device configuration

the observed inefficiency of the XY ratchet may be the neglect of quenched disorder whose account allows one to build a link between muscle architecture and the theory of spin glasses and reveal a tight relation between actomyosin disregistry and the optimal mechanical performance of the force generating machinery [21].

The schematic nature of the proposed models, allowing one to identify only the main mechanisms, is the main reason why we did not attempt to perform a systematic quantitative comparison of our predictions with experiment. Extending the proposed framework towards the account of collectively interacting cross-bridges will open the possibility to calibrate the model using experimental data. Given the purely mechanical nature of our modeling, one can then consider building the actual artificial molecular size devices based on the principles developed in these lecture notes.

The proposed framework also raises some specific issues which need to be addressed in future work. One challenge is to understand the microscopic nature of the hysteretic element and of the active mechanism ensuring the non-potential force structure. Another challenge is to find the optimal interaction of our three active mechanisms ensuring the highest performance of the power-stroke driven motor. The third challenge is to study the effects of short range interaction of elastically coupled power-stroke-driven motors.

The experimental verification of the proposed model of active rigidity requires quantitative monitoring of the mechanical properties of a biological system (say, cytoskeleton) combined with the control of activity elements (say, molecular motors) and the corresponding experimental techniques are already available [1, 37]. The mastery of actively tunable rigidity will open interesting prospects not only in biomechanics [92] but also in engineering design incorporating negative stiffness [39] or aiming at synthetic materials involving dynamic stabilization [16, 101].

References

1. W.W. Ahmed, É. Fodor, T. Betz, Active cell mechanics: measurement and theory. Biochim. Biophys. Acta, Mol. Cell Res. **1853**(11), 3083–3094 (2015)
2. R. Ait-Haddou, W. Herzog, Brownian ratchet models of molecular motors. Cell Biochem. Biophys. **38**(2), 191–213 (2003)
3. B. Alberts, *Molecular Biology of the Cell*, 5th edn. (Garland Science, New York, 2007)
4. F. Alouges, A. DeSimone, A. Lefebvre, Optimal strokes for low Reynolds number swimmers: an example. J. Nonlinear Sci. **18**(3), 277–302 (2008)
5. J. Alvarado, M. Sheinman, A. Sharma, F.C. MacKintosh, G.H. Koenderink, Molecular motors robustly drive active gels to a critically connected state. Nat. Phys. **9**, 591—597 (2013)
6. J.P. Baltanás, L. Lopez, I.I. Blechman, P.S. Landa, A. Zaikin, J. Kurths, M.A.F. Sanjuán, Experimental evidence, numerics, and theory of vibrational resonance in bistable systems. Phys. Rev. E **67**(6), 066119 (2003)
7. C. Barclay, R. Woledge, N. Curtin, Inferring crossbridge properties from skeletal muscle energetics. Prog. Biophys. Mol. Bio. **102**(1), 53–71 (2010)
8. R. Bartussek, Ratchets driven by colored gaussian noise, in *Stochastic Dynamics*, ed. by L. Schimansky-Geier, T. Pöschel. Lecture Notes in Physics, vol. 484 (Springer, Berlin, 1997), pp. 68–80

9. U. Basu, C. Maes, K. Netočný, How statistical forces depend on the thermodynamics and kinetics of driven media. Phys. Rev. Lett. **114**(25), 250601 (2015)
10. I. Bena, C. Van den Broeck, R. Kawai, K. Lindenberg, Drift by dichotomous Markov noise. Phys. Rev. E **68**(4), 041111 (2003)
11. L. Berthier, J. Kurchan, Non-equilibrium glass transitions in driven and active matter. Nat. Phys. **9**(5), 310–314 (2013)
12. J. Bialké, J.T. Siebert, H. Löwen, T. Speck, Negative interfacial tension in phase-separated active Brownian particles. Phys. Rev. Lett. **115**(9), 098301 (2015)
13. I.I. Blekhman, *Vibrational Mechanics: Nonlinear Dynamic Effects, General Approach, Applications* (World Scientific, Singapore, 2000)
14. J.-P. Bouchaud, A. Georges, Anomalous diffusion in disordered media: statistical mechanisms, models and physical applications. Phys. Rep. **195**(4), 127–293 (1990)
15. Z. Bryant, D. Altman, J. Spudich, The power stroke of myosin vi and the basis of reverse directionality. Proc. Nat. Acad. Sci. U.S.A. **104**(3), 772–777 (2007)
16. M. Bukov, L. D'Alessio, A. Polkovnikov, Universal high-frequency behavior of periodically driven systems: from dynamical stabilization to Floquet engineering. Adv. Phys. **64**(2), 139–226 (2015)
17. E.I. Butikov, An improved criterion for Kapitza's pendulum stability. J. Phys. A Math. Theor. **44**(29), 295202 (2011)
18. F.J. Cao, L. Dinis, J.M.R. Parrondo, Feedback control in a collective flashing ratchet. Phys. Rev. Lett. **93**(4), 040603 (2004)
19. M. Caruel, J.-M. Allain, L. Truskinovsky, Muscle is a meta-material operating at a critical point. Phys. Rev. Lett. **110**(24), 248103 (2013)
20. L.F. Cugliandolo, D.R. Grempel, C.A. da Silva Santos, From second to first order transitions in a disordered quantum magnet. Phys. Rev. Lett. **85**(12), 2589–2592 (2000)
21. H.B. da Rocha, L. Truskinovsky, Functionality of disorder in muscle mechanics. Phys. Rev. Lett. **122**(8), 088103 (2019)
22. E. De La Cruz, E. Ostap, Relating biochemistry and function in the myosin superfamily. Curr. Opin. Cell Biol. **16**(1):61–67 (2004)
23. S. Denisov, Particle with internal dynamical asymmetry: chaotic self-propulsion and turning. Phys. Lett. A **296**(4), 197–203 (2002)
24. I. Derényi, T. Vicsek, The Kinesin walk: A dynamic model with elastically coupled heads. Proc. Nat. Acad. Sci. U.S.A. **93**(13), 6775–6779 (1996)
25. C.R. Doering, W. Horsthemke, J. Riordan, Nonequilibrium fluctuation-induced transport. Phys. Rev. Lett. **72**(19), 2984 (1994)
26. T. Duke, Molecular model of muscle contraction. Proc. Nat. Acad. Sci. U.S.A. **96**(6), 2770–2775 (1999)
27. T. Duke, S. Leibler, Motor protein mechanics: a stochastic model with minimal mechanochemical coupling. Biophys. J. **71**(3), 1235–1247 (1996)
28. B. Dybiec, E. Gudowska-Nowak, I.M. Sokolov, Stationary states in Langevin dynamics under asymmetric Lévy noises. Phys. Rev. E **76**(4), 041122 (2007)
29. B. Dybiec, L. Schimansky-Geier, Emergence of bimodality in noisy systems with single-well potential. Eur. Phys. J. B **57**(3), 313–320 (2007)
30. K.A.P. Edman, Double-hyperbolic force velocity relation in frog-muscle fibers. J. Physiol. Lond. **404**, 301–321 (1988)
31. K.A.P. Edman, A. Månsson, C. Caputo, The biphasic force-velocity relationship in frog muscle fibres and its evaluation in terms of cross-bridge function. J. Physiol. **503**(1), 141–156 (1997)
32. T. Elston, C. Peskin, The role of protein flexibility in molecular motor function: coupled diffusion in a tilted periodic potential. SIAM J. Appl. Math. **60**(3), 842–867 (2000)
33. T. Erdmann, U. Schwarz, Stability of adhesion clusters under constant force. Phys. Rev. Lett. **92**(10), 108102 (2004)
34. N. Etemadi, M. Kaminski, Strong law of large numbers for 2-exchangeable random variables. Stat. Probab. Lett. **28**(3), 245–250 (1996)

35. M. Feito, J.P. Baltanás, F.J. Cao, Rocking feedback-controlled ratchets. Phys. Rev. E **80**(3), 031128 (2009)
36. M. Feito, F. J. Cao, Time-delayed feedback control of a flashing ratchet. Phys. Rev. E **76**(6), 061113 (2007)
37. É. Fodor, W.W. Ahmed, M. Almonacid, M. Bussonnier, N.S. Gov, M.-H. Verlhac, T. Betz, P. Visco, F. van Wijland, Nonequilibrium dissipation in living oocytes (2015). arXiv:1511.00921
38. C. Fogle, J. Rudnick, D. Jasnow, Protein viscoelastic dynamics: a model system (2015). arXiv:1502.00343 [cond-mat]
39. F. Fritzen, D.M. Kochmann, Material instability-induced extreme damping in composites: a computational study. Int. J. Solids Struct. **51**(23–24), 4101–4112 (2014)
40. R. Gallardo, O. Idigoras, P. Landeros, A. Berger, Analytical derivation of critical exponents of the dynamic phase transition in the mean-field approximation. Phys. Rev. E **86**(5), 051101 (2012)
41. L. Gammaitoni, P. Hänggi, P. Jung, F. Marchesoni, Stochastic resonance: a remarkable idea that changed our perception of noise. Eur. Phys. J. B **69**(1), 1–3 (2009)
42. M.A. Geeves, Stretching the lever-arm theory. Nature **415**(6868), 129–131 (2002)
43. B. Geislinger, R. Kawai, Brownian molecular motors driven by rotation-translation coupling. Phys. Rev. E **74**(1), 011912 (2006)
44. A. Grosberg, J.-F. Joanny, Nonequilibrium statistical mechanics of mixtures of particles in contact with different thermostats. Phys. Rev. E **92**(3), 032118 (2015)
45. T. Guérin, J. Prost, P. Martin, J.-F. Joanny, Coordination and collective properties of molecular motors: theory. Curr. Opin. Cell Biol. **22**(1), 14–20 (2010)
46. T. Guérin, J. Prost, P. Martin, J.-F. Joanny, Dynamical behavior of molecular motor assemblies in the rigid and crossbridge models. Eur. Phys. J. E **34**(6), 60 (2011)
47. A.N. Gupta, A. Vincent, K. Neupane, H. Yu, F. Wang, M.T. Woodside, Experimental validation of free-energy-landscape reconstruction from non-equilibrium single-molecule force spectroscopy measurements. Nat. Phys. **7**(8), 631–634 (2011)
48. P. Hänggi, P. Jung, Colored noise in dynamical-systems, in *Advances in Chemical Physics*, vol. 89 (Wiley, London, 1995), pp. 239–326
49. P. Hänggi, F. Marchesoni, Artificial Brownian motors: Controlling transport on the nanoscale. Rev. Mod. Phys. **81**(1), 387–442 (2009)
50. D. Hennig, Current control in a tilted washboard potential via time-delayed feedback. Phys. Rev. E **79**(4), 041114 (2009)
51. A.V. Hill, The heat of shortening and the dynamic constants of muscle. Proc. R. Soc. Lond. B **126**, 136–195 (1938)
52. T. Hill, Theoretical formalism for the sliding filament model of contraction of striated muscle. part I. Prog. Biophys. Mol. Biol. **28**, 267–340 (1974)
53. J. Howard, *Mechanics of Motor Proteins and the Cytoskeleton* (Sinauer Associates, Sunderland, 2001)
54. A.F. Huxley, R.M. Simmons, Proposed mechanism of force generation in striated muscle. Nature **233**(5321), 533–538 (1971)
55. A. Ichiki, Y. Tadokoro, M.I. Dykman, Singular response of bistable systems driven by telegraph noise. Phys. Rev. E **85**(3), 031106 (2012)
56. M. Joyeux, E. Bertin, Pressure of a gas of underdamped active dumbbells. Phys. Rev. E **93**(3), 032605 (2016)
57. F. Jülicher, Force and motion generation of molecular motors: a generic description, in *Transport and Structure* (Springer, Berlin, 1999), pp. 46–74
58. F. Jülicher, A. Ajdari, J. Prost, Modeling molecular motors. Rev. Mod. Phys. **69**(4), 1269–1281 (1997)
59. F. Jülicher, J. Prost, Spontaneous oscillations of collective molecular motors. Phys. Rev. Lett. **78**(23), 4510–4513 (1997)
60. K. Kawaguchi, S. Ishiwata, Temperature dependence of force, velocity, and processivity of single Kinesin molecules. Biochem. Bioph. Res. Commun. **272**(3), 895–899 (2000)

61. M. Khoury, J.P. Gleeson, J.M. Sancho, A.M. Lacasta, K. Lindenberg, Diffusion coefficient in periodic and random potentials. Phys. Rev. E **80**(2), 021123 (2009)
62. K. Kitamura, M. Tokunaga, S. Esaki, A. Iwane, T. Yanagida, Mechanism of muscle contraction based on stochastic properties of single actomyosin motors observed in vitro. Biophysics **1**, 1–19 (2005)
63. P. Kloeden, E. Platen, *Numerical Solution of Stochastic Differential Equations*. Applications of Mathematics, vol. 23 (Springer, Berlin, 1992)
64. V. Klyatskin, Dynamic systems with parameter fluctuations of the telegraphic-process type. Radiofizika **20**(4), 562–575 (1977)
65. G. Lan, S.X. Sun, Dynamics of myosin-driven skeletal muscle contraction: I. Steady-state force generation. Biophys. J. **88**(6), 4107 (2005)
66. P.S. Landa, P.V.E. McClintock, Nonlinear systems with fast and slow motions. changes in the probability distribution for fast motions under the influence of slower ones. Phys. Rep. **532**(1), 1–26 (2013)
67. A. Lewalle, W. Steffen, O. Stevenson, Z. Ouyang, J. Sleep, Single-molecule measurement of the stiffness of the Rigor myosin head. Biophys. J. **94**(6), 2160–2169 (2008)
68. Y.-X. Li, Brownian motors possessing internal degree of freedom. Phys. A **251**(3), 382–388 (1998)
69. M. Linari, M. Caremani, V. Lombardi, A kinetic model that explains the effect of inorganic phosphate on the mechanics and energetics of isometric contraction of fast skeletal muscle. Proc. Biol. Sci. **277**(1678), 19–27 (2010)
70. B. Lindner, M. Kostur, L. Schimansky-Geier, Optimal diffusive transport in a tilted periodic potential. Fluct. Noise Lett. **1**(1), R25–R39 (2001)
71. V. Lombardi, G. Piazzesi, The contractile response during steady lengthening of stimulated frog-muscle fibers. J. Physiol. Lond. **431**, 141–171 (1990)
72. R. Lymn, E. Taylor, Mechanism of adenosine triphosphate hydrolysis by actomyosin. Biochemistry **10**(25), 4617–4624 (1971)
73. Magnasco, Molecular combustion motors. Phys. Rev. Lett. **72**(16), 2656–2659 (1994)
74. M.O. Magnasco, Forced thermal ratchets. Phys. Rev. Lett. **71**(10), 1477–1481 (1993)
75. Y.A. Makhnovskii, V.M. Rozenbaum, D.-Y. Yang, S.H. Lin, Net transport due to noise-induced internal reciprocating motion. J. Chem. Phys. **130**(16), 164101 (2009)
76. L. Marcucci, L. Truskinovsky, Mechanics of the power stroke in myosin II. Phys. Rev. E **81**(5), 051915 (2010)
77. P. Martin, A.D. Mehta, A.J. Hudspeth, Negative hair-bundle stiffness betrays a mechanism for mechanical amplification by the hair cell. PNAS **97**(22), 12026–12031 (2000)
78. J.L. Mateos, F. Alatriste, Brownian motors and stochastic resonance. Chaos **21**(4) (2011)
79. J. Menche, L. Schimansky-Geier, Two particles with bistable coupling on a ratchet. Phys. Lett. A **359**(2), 90–98 (2006)
80. E. Meyhöfer, J. Howard, The force generated by a single Kinesin molecule against an elastic load. Proc. Nat. Acad. Sci. U.S.A. **92**(2), 574–578 (1995)
81. M.M. Millonas, M.I. Dykman, Transport and current reversal in stochastically driven ratchets. Phys. Lett. A **185**(1), 65–69 (1994)
82. A. Mogilner, A.J. Fisher, R.J. Baskin, Structural changes in the neck linker of Kinesin explain the load dependence of the motor's mechanical cycle. J. Theor. Biol. **211**(2), 143–157 (2001)
83. A. Månsson, Actomyosin-ADP states, interhead cooperativity, and the force-velocity relation of skeletal muscle. Biophys. J. **98**(7), 1237–1246 (2010)
84. A. Muhlrad, Y.M. Peyser, M. Nili, K. Ajtai, E. Reisler, T.P. Burghardt, Chemical decoupling of ATPase activation and force production from the contractile cycle in myosin by steric hindrance of lever-arm movement. Biophys. J. **84**(2), 1047 (2003)
85. M.A. Muñoz, F.D.L. Santos, M.M.T.D. Gama, Generic two-phase coexistence in nonequilibrium systems. Eur. Phys. J. B **43**(1):73–79 (2005)
86. K.H. Nagai, Y. Sumino, R. Montagne, I.S. Aranson, H. Chaté, Collective motion of self-propelled particles with memory. Phys. Rev. Lett. **114**(16), 168001 (2015)

87. E. Pate, G. Wilson, M. Bhimani, R. Cooke, Temperature-dependence of the inhibitory of effects on orthovanadate on shortening velocity in fast skeletalimuscle. Biophys. J. **66**(5), 1554–1562 (1994)

88. G. Piazzesi, V. Lombardi, A cross-bridge model that is able to explain mechanical and energetic properties of shortening muscle. Biophys. J. **68**(5), 1966–1979 (1995)

89. K.R. Pilkiewicz, J.D. Eaves, Reentrance in an active glass mixture. Soft Matter **10**(38), 7495–7501 (2014)

90. M. Porto, Molecular motor based entirely on the coulomb interaction. Phys. Rev. E **63**(3), 030102 (2001)

91. J. Prost, J.-F. Chauwin, L. Peliti, A. Ajdari, Asymmetric pumping of particles. Phys. Rev. Lett. **72**(16), 2652 (1994)

92. G. Puglisi, L. Truskinovsky, Cohesion-decohesion asymmetry in geckos. Phys. Rev. E **87**(3), 032714 (2013)

93. H. Qian, Vector field formalism and analysis for a class of thermal ratchets. Phys. Rev. Lett. **81**(15), 3063–3066 (1998)

94. P. Recho, L. Truskinovsky, Asymmetry between pushing and pulling for crawling cells. Phys. Rev. E **87**(2), 022720 (2013)

95. P. Reimann, Brownian motors: noisy transport far from equilibrium. Phys. Rep. **361**(2–4), 57–265 (2002)

96. P. Reimann, Brownian motors: noisy transport far from equilibrium. Phys. Rep. **361**(2–4), 57–265 (2002)

97. P. Reimann, C. Van den Broeck, H. Linke, P. Hänggi, J.M. Rubi, A. Pérez-Madrid, Giant acceleration of free diffusion by use of tilted periodic potentials. Phys. Rev. Lett. **87**(1), 010602 (2001)

98. H. Risken, *The Fokker–Planck Equation* (Springer, Berlin, 1989)

99. V.M. Rozenbaum, Y.A. Makhnovskii, S.-Y. Sheu, D.-Y. Yang, S.H. Lin, Two-state brownian motor driven by synchronously fluctuating unbiased forces. Phys. Rev. E **84**(2), 021104 (2011)

100. P. Sarkar, A.K. Maity, A. Shit, S. Chattopadhyay, J.R. Chaudhuri, S.K. Banik, Controlling mobility via rapidly oscillating time-periodic stimulus. Chem. Phys. Lett. **602**, 4–9 (2014)

101. P. Sarkar, A. Shit, S. Chattopadhyay, S. K. Banik, Profiling the overdamped dynamics of a nonadiabatic system. Chem. Phys. **458**, 86–91 (2015)

102. S. Savel'ev, F. Marchesoni, F. Nori, Stochastic transport of interacting particles in periodically driven ratchets. Phys. Rev. E **70**(6), 061107 (2004)

103. G. Schappacher-Tilp, T. Leonard, G. Desch, W. Herzog, A novel three-filament model of force generation in eccentric contraction of skeletal muscles. PLoS One **10**(3), e0117634 (2015)

104. M.J. Schnitzer, K. Visscher, S.M. Block, Force production by single Kinesin motors. Nat. Cell Biol. **2**(10), 718–723 (2000)

105. K. Sekimoto, Kinetic characterization of heat bath and the energetics of thermal ratchet models. J. Phys. Soc. Jpn. **66**(5), 1234–1237 (1997)

106. M. Sheinman, C.P. Broedersz, F.C. MacKintosh, Actively stressed marginal networks. Phys. Rev. Lett. **109**(23), 238101 (2012)

107. R. Sheshka, P. Recho, L. Truskinovsky, Rigidity generation by nonthermal fluctuations. Phys. Rev. E **93**(5), 052604 (2016)

108. R. Sheshka, L. Truskinovsky, Power-stroke-driven actomyosin contractility. Phys. Rev. E **89**(1), 012708 (2014)

109. D.A. Smith, M.A. Geeves, J. Sleep, S.M. Mijailovich, Towards a unified theory of muscle contraction. I: foundations. Ann. Biomed. Eng. **36**(10), 1624–1640 (2008)

110. A.P. Solon, J. Stenhammar, R. Wittkowski, M. Kardar, Y. Kafri, M.E. Cates, J. Tailleur, Pressure and phase equilibria in interacting active Brownian spheres. Phys. Rev. Lett. **114**(19), 198301 (2015)

111. H. Sugi, T. Kobayashi, T. Tsuchiya, S. Chaen, S. Sugiura, Evidence for the essential role of myosin head lever arm domain and myosin subfragment-2 in muscle contraction, in Skeletal Muscle—From Myogenesis to Clinical Relations, ed. by J. Cseri, chap. 6 (InTech, 2012), pp. 125–140

112. H. Sugi, H. Minoda, Y. Inayoshi, F. Yumoto, T. Miyakawa, Y. Miyauchi, M. Tanokura, T. Akimoto, T. Kobayashi, S. Chaen, et al., Direct demonstration of the cross-bridge recovery stroke in muscle thick filaments in aqueous solution by using the hydration chamber. Proc. Nat. Acad. Sci. U.S.A. **105**(45), 17396–17401 (2008)

113. H.L. Sweeney, A. Houdusse, Structural and functional insights into the myosin motor mechanism. Annu. Rev. Biophys. **39**, 539–557 (2010)

114. S.C. Takatori, J.F. Brady, Forces, stresses and the (thermo?) dynamics of active matter. Curr. Opin. Colloid Interface Sci. **21**, 24–33 (2016)

115. T. Tomé, M.J. de Oliveira, Dynamic phase transition in the kinetic Ising model under a time-dependent oscillating field. Phys. Rev. A **41**(8), 4251 (1990)

116. G. Tsiavaliaris, S. Fujita-Becker, D.J. Manstein, Molecular engineering of a backwards-moving myosin motor. Nature **427**(6974), 558–561 (2004)

117. M. Tyska, D. Warshaw, The myosin power stroke. Cell Mot. Cytoskel. **51**(1), 1–15 (2002)

118. C. Van den Broeck, J.M.R. Parrondo, R. Toral, Noise-induced nonequilibrium phase transition. Phys. Rev. Lett. **73**(25), 3395 (1994)

119. C. Veigel, C.F. Schmidt, Moving into the cell: single-molecule studies of molecular motors in complex environments. Nat. Rev. Mol. Cell Bio. **12**(3), 163–176 (2011)

120. A. Vilfan, T. Duke, Instabilities in the transient response of muscle. *Biophys. J.* **85**(2), 818–827 (2003)

121. A. Vilfan, T. Duke, Synchronization of active mechanical oscillators by an inertial load. Phys. Rev. Lett. **91**(11), 114101 (2003)

122. A. Vilfan, E. Frey, F. Schwabl, Force-velocity relations of a two-state crossbridge model for molecular motors. Europhys. Lett. **45**(3), 283–289 (1999)

123. K. Visscher, M.J. Schnitzer, S.M. Block, Single Kinesin molecules studied with a molecular force clamp. Nature **400**(6740), 184–189 (1999)

124. S. von Gehlen, M. Evstigneev, P. Reimann, Dynamics of a dimer in a symmetric potential: Ratchet effect generated by an internal degree of freedom. Phys. Rev. E **77**(3), 031136 (2008)

125. S. Walcott, P. Fagnant, K. Trybus, D. Warshaw, Smooth muscle heavy meromyosin phosphorylated on one of its two heads supports force and motion. J. Biol. Chem. **284**(27), 18244–18251 (2009)

126. H. Wang, G. Oster, Ratchets, power strokes, and molecular motors. Appl. Phys. A **75**(2), 315–323 (2002)

127. H. J. Woo, C.L. Moss, Analytical theory of the stochastic dynamics of the power stroke in nonprocessive motor proteins. Phys. Rev. E **72**(5), 051924 (2005)

128. M.T. Woodside, C. García-García, S.M. Block, Folding and unfolding single rna molecules under tension. Curr. Opin. Chem. Biol. **12**(6), 640–646 (2008)

129. A.A. Zaikin, J. Kurths, L. Schimansky-Geier, Doubly stochastic resonance. Phys. Rev. Lett. **85**(2), 227 (2000)

130. X.-J. Zhang, H. Qian, M. Qian, Stochastic theory of nonequilibrium steady states and its applications. Part I. Phys. Rep. **510**(1), 1–86 (2012)

LECTURE NOTES IN MATHEMATICS

 Springer

Editors in Chief: J.-M. Morel, B. Teissier;

Editorial Policy

1. Lecture Notes aim to report new developments in all areas of mathematics and their applications – quickly, informally and at a high level. Mathematical texts analysing new developments in modelling and numerical simulation are welcome.

   Manuscripts should be reasonably self-contained and rounded off. Thus they may, and often will, present not only results of the author but also related work by other people. They may be based on specialised lecture courses. Furthermore, the manuscripts should provide sufficient motivation, examples and applications. This clearly distinguishes Lecture Notes from journal articles or technical reports which normally are very concise. Articles intended for a journal but too long to be accepted by most journals, usually do not have this "lecture notes" character. For similar reasons it is unusual for doctoral theses to be accepted for the Lecture Notes series, though habilitation theses may be appropriate.

2. Besides monographs, multi-author manuscripts resulting from SUMMER SCHOOLS or similar INTENSIVE COURSES are welcome, provided their objective was held to present an active mathematical topic to an audience at the beginning or intermediate graduate level (a list of participants should be provided).

   The resulting manuscript should not be just a collection of course notes, but should require advance planning and coordination among the main lecturers. The subject matter should dictate the structure of the book. This structure should be motivated and explained in a scientific introduction, and the notation, references, index and formulation of results should be, if possible, unified by the editors. Each contribution should have an abstract and an introduction referring to the other contributions. In other words, more preparatory work must go into a multi-authored volume than simply assembling a disparate collection of papers, communicated at the event.

3. Manuscripts should be submitted either online at www.editorialmanager.com/lnm to Springer's mathematics editorial in Heidelberg, or electronically to one of the series editors. Authors should be aware that incomplete or insufficiently close-to-final manuscripts almost always result in longer refereeing times and nevertheless unclear referees' recommendations, making further refereeing of a final draft necessary. The strict minimum amount of material that will be considered should include a detailed outline describing the planned contents of each chapter, a bibliography and several sample chapters. Parallel submission of a manuscript to another publisher while under consideration for LNM is not acceptable and can lead to rejection.

4. In general, **monographs** will be sent out to at least 2 external referees for evaluation.

   A final decision to publish can be made only on the basis of the complete manuscript, however a refereeing process leading to a preliminary decision can be based on a pre-final or incomplete manuscript.

   Volume Editors of **multi-author works** are expected to arrange for the refereeing, to the usual scientific standards, of the individual contributions. If the resulting reports can be forwarded to the LNM Editorial Board, this is very helpful. If no reports are forwarded or if other questions remain unclear in respect of homogeneity etc, the series editors may wish to consult external referees for an overall evaluation of the volume.

5. Manuscripts should in general be submitted in English. Final manuscripts should contain at least 100 pages of mathematical text and should always include

   – a table of contents;
   – an informative introduction, with adequate motivation and perhaps some historical remarks: it should be accessible to a reader not intimately familiar with the topic treated;
   – a subject index: as a rule this is genuinely helpful for the reader.
   – For evaluation purposes, manuscripts should be submitted as pdf files.

6. Careful preparation of the manuscripts will help keep production time short besides ensuring satisfactory appearance of the finished book in print and online. After acceptance of the manuscript authors will be asked to prepare the final LaTeX source files (see LaTeX templates online: https://www.springer.com/gb/authors-editors/book-authors-editors/manuscriptpreparation/5636) plus the corresponding pdf- or zipped ps-file. The LaTeX source files are essential for producing the full-text online version of the book, see http://link.springer.com/bookseries/304 for the existing online volumes of LNM). The technical production of a Lecture Notes volume takes approximately 12 weeks. Additional instructions, if necessary, are available on request from lnm@springer.com.

7. Authors receive a total of 30 free copies of their volume and free access to their book on SpringerLink, but no royalties. They are entitled to a discount of 33.3 % on the price of Springer books purchased for their personal use, if ordering directly from Springer.

8. Commitment to publish is made by a *Publishing Agreement*; contributing authors of multiauthor books are requested to sign a *Consent to Publish form*. Springer-Verlag registers the copyright for each volume. Authors are free to reuse material contained in their LNM volumes in later publications: a brief written (or e-mail) request for formal permission is sufficient.

Addresses:
Professor Jean-Michel Morel, CMLA, École Normale Supérieure de Cachan, France
E-mail: moreljeanmichel@gmail.com

Professor Bernard Teissier, Equipe Géométrie et Dynamique,
Institut de Mathématiques de Jussieu – Paris Rive Gauche, Paris, France
E-mail: bernard.teissier@imj-prg.fr

Springer: Ute McCrory, Mathematics, Heidelberg, Germany,
E-mail: lnm@springer.com

Printed in the United States
By Bookmasters